# Florida Manatees

## BIOLOGY, BEHAVIOR, AND CONSERVATION

John E. Reynolds III
Photographs by Wayne Lynch

JOHNS HOPKINS UNIVERSITY PRESS · BALTIMORE

Johns Hopkins University Press
2715 North Charles Street
Baltimore, Maryland 21218-4363
www.press.jhu.edu

Library of Congress Cataloging-in-Publication Data

Names: Reynolds, John Elliott, 1952– , author. | Lynch,
    Wayne, illustrator.
Title: Florida manatees : biology, behavior, and conservation
    / John E. Reynolds ; photographs by Wayne Lynch.
Description: Baltimore : Johns Hopkins University Press,
    2017. | Includes bibliographical references and index.
Identifiers: LCCN 2016022349 | ISBN 9781421421919
    (hardcover : alk. paper) | ISBN 9781421421926 (elec-
    tronic) | ISBN 1421421917 (hardcover : alk. paper) | ISBN
    1421421925 (electronic)
Subjects: LCSH: West Indian manatee. | West Indian mana-
    tee—Pictorial works.
Classification: LCC QL737.S63 R486 2017 | DDC 599.55—
    dc23
    LC record available at https://lccn.loc.gov/2016022349

A catalog record for this book is available from the
British Library.

*Special discounts are available for bulk purchases of this book.
For more information, please contact Special Sales at 410-516-
6936 or specialsales@press.jhu.edu.*

Johns Hopkins University Press uses environmentally friendly
book materials, including recycled text paper that is composed of at least 30 percent post-consumer waste, whenever
possible.

All photographs © Wayne Lynch except as noted:

Page 2: Guadaloupe Island map inset, © Eric Gaba, Wikime-
dia Commons

Pages 2, 4, 32: Background maps, © www.free-world-maps
.com

Page 51: Inset of manatees huddling near a power plant,
© John Reynolds

Page 63: Composite of photographer Wayne Lynch, © Kathleen Parker

# Contents

# Acknowledgments

*From Wayne Lynch*  This is my fourth book with Johns Hopkins University Press, and, as always, the process has been an enjoyable one, made all the more so by science editor Vincent J. Burke. I thank him for his professional guidance and friendly, pleasant manner. I also wish to thank Robert Bonde with the US Geological Survey Sirenia Project, who invited me to join him on several manatee captures and allowed me to examine and photograph his impressive array of sirenian collectibles. Others who helped with the project include Buddy Powell with the Sea to Shore Alliance, who used his private plane to give me an overview of coastal manatee habitat; Ivan Vicente, a wildlife biologist with the Crystal River Wildlife Refuge, who arranged for me to get photo permits year after year; and Susan and Mike Kirk, who had manatees in their backyard and generously gave me access to their property whenever I wanted. I reserve special thanks for Joree and David Cox, who loaned me, a total stranger, a kayak to get close to the manatees that swam by their house, and they did so countless times over four years. My photo coverage of manatees would never have been as extensive without their friendly help. It was an honor to collaborate with John Reynolds, a highly respected scientist and author. Before joining up with this project, I had enjoyed all of his previous books on marine mammals, and they were a valued part of my personal library.

Finally, I wish to thank my partner and wife of forty-one years, Aubrey Lang, who was with me every day of the project. We swam together with manatees innumerable times and always discussed our outings over hot coffee as we tried to warm our bodies. Throughout, she was a constant source of encouragement, enthusiasm, and intelligent suggestions. I love her dearly.

*From John E. Reynolds III*  First, I thank Johns Hopkins University Press editor Vincent J. Burke for inviting me to contribute some text to accompany the spectacular images provided by Wayne Lynch. I am, of course, grateful to Wayne for enthusiastically welcoming me to his team; I am honored to collaborate with him on this book, and I am confident that interest in the end product will be due primarily to his beautiful photographs, rather than to my modest text.

I am grateful to a number of other individuals, besides Vince Burke, associated with Johns Hopkins University Press. I especially want to recognize editorial assistant Meagan Szekely and copyeditor Kathleen Capels. Their painstaking efforts greatly improved the quality of the text.

I also thank the members of my staff of the Manatee Research Program at Mote Marine Laboratory. They (alphabetically) are Sheri Barton, Anne Honeywell, Jennifer Johnson, and Kerri Scolardi. They reviewed drafts of the text and offered excellent, constructive criticism. In this and other ventures, I have appreciated their knowledge, passion, friendship, and dedication to manatee research and conservation.

My introduction to manatees occurred when I started graduate school at the University of Miami in 1974. I have been privileged to know and work with most of the leading sirenian biologists and conservationists (indeed, with many of the leading marine mammal scientists and conservationists) in the world since that time. I am grateful to all of them for helping me learn and for their friendship. As I have stated many times, I should have been the poster child for the saying, "I'd rather be lucky than good!"

Finally, and most important, I thank my family (including my new grandson, Jack) and friends for their support and love.

Although wild manatees are now absent from Guadeloupe waters, some other large marine vertebrates call the islands home. Ponderous leatherback turtles come ashore to nest, and the waters offshore have giant sperm whales and other marine mammals.

# 1

# *From a Bluff on Guadeloupe*

I am back on the island of Guadeloupe, part of the French West Indies, for perhaps the tenth or twelfth time. Even though my mind runs with thoughts of manatees, there are none to be seen. The reason I am on Guadeloupe is because the manatees that were there are long gone, wiped out almost a century ago. Now some visionary French ecologists and conservationists want them back, and they think my familiarity with Florida manatee biology and conservation might be useful. When on Guadeloupe, I generally stay on the southern side of the flatter, more easterly island of Grand-Terre. In my downtime, I seek the high battlements from which I can catch a cooling sea breeze as I look out over the clear waters and think of the not-too-distant time when manatees hopefully will reoccupy Guadeloupe—and I think I am very lucky to be a marine mammal biologist and conservationist at the start of the twenty-first century.

The challenges to effective conservation of marine mammals are myriad and daunting, but thanks to new tools and some daring and creative ventures, the chances for conservation of manatees and other charismatic marine mammals are better in some ways than they were when I was a child. The story of Guadeloupe's manatees underscores some of the challenges to conservation, the triumphs of a group of de-

Guadeloupe, a territory of France, is an archipelago of several islands located in the Leeward Islands of the Lesser Antilles. The two largest islands are Basse-Terre and Grande-Terre. Folklore indicates that manatees were hunted to extinction in Guadeloupe about 100 years ago; now the government of France and the Parc National de la Guadeloupe want to reintroduce the species to local waters.

termined people, and how lucky my fellow Floridians are to have several thousand manatees nearby.

## The Caribbean Saga and the Triumph of Guadeloupe

Daryl Domning of Howard University is the unchallenged dean of the study of sirenian (manatee and dugong) fossils and evolution. In one of his many publications he writes that, even though only four species of sirenians exist today, their diversity in the Caribbean was much greater in the distant past. The Caribbean Sea and the adjoining part of the Atlantic Ocean represent the only area on Earth continuously occupied by sirenians since the Eocene epoch, which ended 33.9 million years ago. Sirenian diversity in the Caribbean peaked in the Oligocene and Miocene epochs (33.9–5.3 million years ago), with four different families (a taxonomic—i.e., scientific—classification for related forms of animals) and a wide range of species represented. In the timeless struggle we associate with natural selection, Caribbean manatees eventually out-competed Caribbean dugongs and other early (and now-extinct) types

of sirenians and became established in the region. Approximately 1–2 million years ago, the modern species called the West Indian manatee (*Trichechus manatus*) came to occupy much of the present range of the species—basically, coastal and river-related waters extending from the southeastern United States to northeastern Brazil.

Of the two taxonomically recognized subspecies of West Indian manatees, the Florida manatee (*Trichechus manatus latirostris*) is found in the waters of the southeastern United States, with occasional wanderers appearing in nearby Cuba and the Bahamas. Some even appear in the waters of mid-Atlantic and New England states and as far west as Texas. The Antillean manatee (*Trichechus manatus manatus*) occupies the remainder of the species' range.

Although the geographic range of West Indian manatees in the Caribbean is very broad, the species is found discontinuously today, occupying waters of some islands and certain Central, South, and North American coastal regions. Overall, the species exists in the waters around almost two dozen countries or territories. Although nobody knows exactly how many West Indian manatees there currently are, scientists guess that they number around 5,500 individuals belonging to the Antillean subspecies and perhaps 6,500 in the Florida subspecies. There are fewer than 100 individuals in many of the Caribbean locations that have manatees present at all, with Puerto Rico, Mexico, Belize, Colombia, and Brazil having the largest populations outside of the United States.

With minimal enforcement of protective laws and the animals' survival coming under pressure from deliberate hunting, accidental en-

It is not just the waters of Guadeloupe that are filled with life. High in the air above the island, seabirds, such as the magnificent frigate bird, are beautiful as they circle, silhouetted against the tropical blue sky. Only male frigate birds have a red gular (throat) sac, which is inflated to attract a mate.

**Florida Manatee**
(*Trichechus manatus latirostris*)

**Antillean Manatee**
(*Trichechus manatus manatus*)

The Florida subspecies of the West Indian manatee is found primarily in coastal, estuarine (areas of brackish water where a river flows into the sea), and riverine waters of the southeastern United States, as well as occasionally in the Bahamas and even Cuba. The range of the Antillean subspecies is far wider. Despite the difference in the size of the subspecies' ranges, the overall numbers of Florida manatees and Antillean manatees are not all that different.

tanglements in nets, pollution, habitat destruction, and collisions with motorboats, many manatee populations in the Caribbean are unlikely to persist very long into the future. Given the small numbers of manatees in most locations, coupled with the breadth and severity of threats to them, I fear that the number of range states (countries or territories in which manatees can be found) could drop from nearly two dozen at present to fewer than half that number in a matter of years.

Guadeloupe, unfortunately, led the way in the early twentieth century in exterminating its manatees, mostly, it appears, for food for the local people. Visitors to Guadeloupe today can spend time in Lamentin, a lovely coastal community on the northern shore of Basse-Terre, the more mountainous of Guadeloupe's two main islands. The town's name comes from the French term for manatee: *lamantin*. The town of Vieux Borg, on the shore of Grande-Terre, was the site of a manatee processing operation, where meat and fat were removed from harvested animals and consumed by the area's residents. Wild manatees may no longer exist in the waters of Guadeloupe, but marks of their former presence are still evident today.

In the early part of the twenty-first century, the government of France seriously began to consider the idea of reintroducing manatees to the waters of Guadeloupe, specifically to a beautiful and sparsely populated embayment (bay-like area) called the Grand Cul-de-Sac Marin, located to the north of the isthmus between Grande-Terre and Basse-Terre. The idea was spawned and nurtured by some local champions of these animals,

most notably Hervé Magnin, Maurice Anselme, and Boris Lerebours of the Parc National de la Guadeloupe (Guadeloupe National Park), which is a component of a global network of ten national parks under the jurisdiction of France and collectively called Parcs Nationaux de France. In 2008, a colleague (toxicologist Dana Wetzel) and I were invited to Guadeloupe by Hervé and the Parc National de la Guadeloupe to provide an assessment of the suitability of the Grande Cul-de-Sac Marin habitat for and possible effects of human activities within it on the introduction of a founder group (small initial population) of manatees. Although we felt that there were certain critical questions to be answered before embarking on such a venture, our opinion was that this area had the potential to provide an outstanding opportunity for manatee reintroduction and regional conservation.

This bold idea, which would be only the second reintroduction ever of a marine mammal to a region from which it was completely exterminated (the first being sea otters on the West Coast of the United States),

Although manatees may not be as beautiful as some other creatures, their placid and unaggressive manner has given them many fans worldwide.

Florida manatees are amazingly graceful swimmers. It is easy to see that this individual is a male, due to the placement of its ventral (abdominal) features, with the one closest to the head (the umbilical scar, or belly button) lying next to the urogenital opening (for reproduction and urine excretion). In females, the umbilical scar is more separated, with the urogenital opening and the anus located more toward the fluke (tail lobe).

has been painstakingly nurtured. Although reintroductions are a standard and important part of the "toolbox" for scientists and managers of terrestrial mammals and some other taxa (scientific groups or entities, such as birds, fish, plants, and the like; the singular form of this word is taxon), the approach has largely been ignored to date for marine mammals. The successful reintroduction of manatees to Guadeloupe could serve as an instructive model to enhance conservation of both manatees and certain other marine mammals in the future.

The Parc National de la Guadeloupe created an Expert Working Group that included recognized specialists in manatee biology, conservation, and veterinary medicine and strove to develop partnerships with fishing groups and other stakeholders that might otherwise resist having manatees back in the area. It now appears that the first manatees to inhabit the waters of Guadeloupe since the past century may be acquired and released in the near future. Adopting the truism that "success begets success," the members of the Expert Working Group expect that an initial, successful introduction of manatees will make subsequent acquisitions and introductions of manatees in Guadeloupe much easier. If the reintroduction succeeds, as I believe it will, not only will it enhance local biodiversity (a goal of all of the Parcs Nationaux de France), but it would also spark the creation of what could eventually become one of the larger and more stable manatee populations in the wider Caribbean, thanks to three factors: (1) existing habitat protection measures by the Parc National de la Guadeloupe, (2) the carrying capacity (the maximum population of a species that an environment can support

A mother manatee is accompanied by her calf, with the latter occupying a common position as the pair swims, the baby staying parallel to its mother, with its head being slightly in back of the mom's.

without undergoing deterioration) for manatees in that habitat, and (3) the endorsement of the local people. In Guadeloupe, manatees are no longer considered to be food; instead, they are beginning to be broadly perceived as living resources with considerable cultural value. A series of well-planned baby steps has the potential to allow Guadeloupe to establish and sustain a manatee population of regional importance.

## Fortunate Floridians

Today there are three species of manatees found worldwide: the Amazonian manatee (*Trichechus inunguis*), the West African manatee (*Trichechus senegalensis*), and the West Indian manatee. The dugong (*Dugong dugon*) is the fourth and final extant species of the taxonomic order Sirenia; the astounding range of dugongs covers the waters of more than three dozen countries and territories in the Indo-Pacific.

In terms of their potential for extinction, all species in the order Sirenia are considered Vulnerable by the International Union for the Conservation of Nature (IUCN), and all are listed in Appendix I (the level

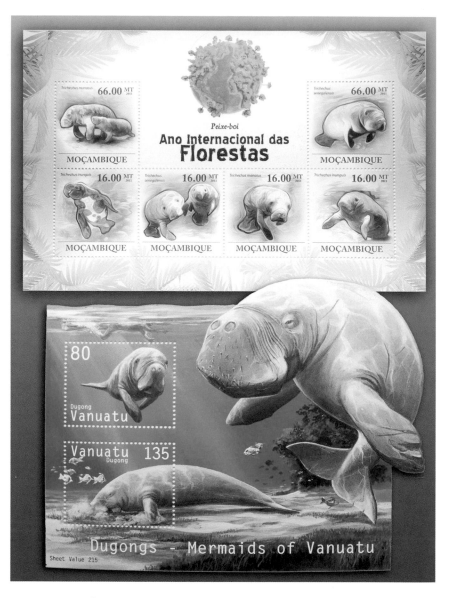

Manatees and dugongs are a prominent part of the art and folklore in countries near where they live. The set of stamps from Mozambique (*top*) depicts all three living species of manatees, whereas the stamps from the Republic of Vanuatu (*bottom*) are of dugongs. Mozambique is not a part of the range for any species of manatee, yet it does have a small but important population of dugongs. The Mozambique stamps were designed to celebrate sirenians and the International Forest Year in Brazil.

representing the most precarious conservation status) by the Convention on International Trade in Endangered Species of Wild Flora and Fauna (CITES). Some dugong populations are in danger of being locally eliminated, but globally there are probably more than 100,000 dugongs, with Australia and the Arabian Gulf being particularly important strongholds. Even given the long-standing needs to study and conserve manatees, there are no reliable population estimates for Amazonian manatees and West African manatees. West African manatees, in particular, seem vulnerable not only to being wiped out from local parts of their range, but also to extinction as a species.

In contrast to the other species of manatees (but not dugongs), West Indian manatees—specifically, Florida manatees—have been well studied for more than forty years by a variety of individuals, private organizations, and agencies. The Florida manatee is unquestionably one of the best-studied marine mammals in the world. Their population size and its trend have been assessed for decades, and the highest single-day count

(6,250 animals) came from a statewide aerial survey of Florida manatees conducted in February 2016. Although the impressive amount of available knowledge regarding the biology and conservation of Florida manatees can often be applied to the relatively unstudied Antillean subspecies, this is not the case when it comes to estimates of their abundance. Based to a large extent on anecdotal information and guesstimates, the number of Antillean manatees across this species' sizeable range is thought to be around 5,500 individuals. Added together, it seems fair to suggest that there were probably about 12,000 West Indian manatees alive in 2015.

With my colleagues Kerri Scolardi and Jane Provancha (and with support from the Florida Power & Light Company), I have conducted several years of aerial surveys to count manatees and plot their distribution and winter habitat use in the waters of the Indian River Lagoon in Brevard County, Florida. We occasionally counted nearly 2,000 manatees in a single day's survey, and we did not even cover all of the waterways in the area, as aerial surveys of some locations are restricted, due to

A manatee comes to the surface to breathe, while other members of the group continue to rest on the bottom. Although the photographs used in this book were taken in clear water, manatees often occupy turbid waters, where visibility is reduced. In such cases, the first sign a person has that manatees are around may be the sound of an explosive exhalation or the sight of a couple of round nostrils just above the surface of the water.

Although people rightly do not consider manatees to be as social as some other species, including the common bottlenose dolphin, manatees in the wild are generally found more often in groups than alone. As is the case for bottlenose dolphins, the composition of those groups may change over the course of a day (or longer time periods) as members break away and rejoin other individuals. The social behavior of manatees has been well studied, because scientists can identify individual animals by their body scars or their mutilated flukes.

proximity to the National Aeronautics and Space Administration's Kennedy Space Center base. Furthermore, this number does not include all of the transient manatees that travel up and down the east coast of the state seasonally. Thus it seems eminently likely that the waters of this one Florida county provide habitat at some point in the year for more than 2,000 individual manatees, accounting for an astounding 30 percent (or more) of the Florida subspecies and equaling something on the order of 15 percent of the West Indian manatees alive in the world today. Occasionally, on cold winter days, we were lucky enough to photograph aggregations (temperature-induced groupings) of manatees at the warm-water refuge created by the discharge from the Florida Power & Light Company's local power plant, called the Cape Canaveral Next Generation Clean Energy Center; in some cases, we captured more than 1,100 manatees in a single image. What a privilege it is to be able to see such a large percentage of a global population of an endangered species, and what an attendant responsibility for Floridians to be caretakers of this resource.

Despite the large numbers that are present, even in the Indian River Lagoon the conservation of manatees is not assured. In Brevard County waterways, recent unusual mortality events, combined with routine regular causes of death, claimed the lives of 704 manatees between 2010 and 2014, as well as possibly an additional 405 manatees in 2015. In that six-year time span, two of the unusual mortality events were traced to cold stress due to severe winter weather, whereas the cause of the third event remains undiagnosed to this day.

It is fairly easy to document manatee mortality in Brevard County, and in some cases to establish the cause of death. Nonetheless, as useful as a body count may be to understand threats to manatees, other components also provide insights with regard to factors that place manatees at risk. For example, significant algal blooms (rapid increases in the quantity of algae that is present in an area) and declines in the amount of seagrasses (which provide forage for manatees) have occurred in Brevard County waters of the Indian River Lagoon in recent years. Seagrass coverage has only started to rebound recently, with a 12 percent increase following a 43 percent loss since 2011. In 2011 alone, nearly 50,000 acres of seagrasses were lost in this county's portion of the Indian River Lagoon. Two decades ago, a symposium addressed the vulnerability of the Indian River Lagoon by promoting improved management specifically designed to protect biodiversity. Threats identified at that time included habitat destruction, watershed alterations, a decline in water quality, and a lack of coordination among management agencies. Those threats remain today, but they are compounded by organized efforts by some

Seagrasses form an important part of the diet of Florida manatees.

The bodies of manatees frequently have algae, barnacles, and other living organisms growing on them. Newborn calves have clean skin, but it does not take long for a youngster to acquire its own community of these freeloaders.

stakeholders to challenge agency regulations and restrictions on their activities. In what is arguably the most important finite body of water in the world for manatees—the Indian River Lagoon—conservation efforts must ensure that habitat quality improves, direct impacts on manatees are minimized, and a comprehensive and proactive conservation ethic is broadly embraced.

With every conservation battle that is waged, good scientific research significantly aids the decisions that must be made. Thus the vast amount of data regarding the behavior, ecology, life history, and physiology of Florida manatees increases the likelihood that scientists and managers will have the ability to conserve West Indian manatees throughout the species' range. This information could assist the citizens, scientists, and managers of Guadeloupe in their efforts to introduce and maintain a viable new population of manatees in the Caribbean. Ultimately, though, success in conservation occurs primarily due to changes in human values and actions, not simply because of the availability of better data. Whether the battle is to sustain the habitat needed by manatees along the east coast of Florida or to establish and nurture a founder population in Guadeloupe, success requires the cooperation of humans who cherish the natural world and the residents thereof.

Descending from the battlements on Grand-Terre following my reflections about manatees, conservation, human values, and how to achieve a balanced approach to sustain both nature and people, I thought more about how fortunate I am to be involved in these important processes.

As Forrest Gump famously stated, "I am not a smart man," but I have been lucky enough to have had some amazing opportunities to study and conserve manatees, and I have been aware enough to make the most of them. The following chapters are my effort to share what I have learned, describe these intriguing animals, and invite you to join the efforts that might help manatees survive and thrive into the future.

**This resting manatee provides the classic, iconic picture of a manatee—massive size and a gentle demeanor, seemingly at peace in a less-than-peaceful world.**

Manatees have abundant, nerve-filled hairs and bristles around their mouths and faces. These are similar to the vibrissae (whiskers) of some other species.

# 2

# A Hodge-Podge of Adaptations

One of the first documentaries to be produced about Florida manatees was entitled *Silent Sirens: Manatees in Peril*; the narrator was the late Leonard Nimoy (famous as Mr. Spock of *Star Trek*), who humorously introduced the creature as "a bewhiskered blimp of an animal." For hundreds of years, the highly unusual appearance and very specialized ecological niche of manatees, coupled with their penchant for living in close proximity to people, have attracted the attention of scientists, wildlife managers, and anyone who cares about nature.

Some of the manatees' adaptations actually work against them in today's world, making them vulnerable to injury, disease, or even death in the coastal waters they occupy. For example, simply being mammals means that manatees must surface every few minutes to breathe, and such occasions can put them at risk from hunters and collisions with watercraft. As herbivores (animals that only eat vegetation), living in an inshore environment where aquatic plants are plentiful can place manatees in areas with high levels of human activity, with its consequent chemical pollution, noise pollution, and physical disturbances. In this chapter, I describe the cadre of traits that make a manatee a manatee, as well as what the implications of those traits are for conservation and management.

There are several taxonomic groups of marine mammals in the world today. This composite contains some representative species: a southern right whale (*top left*); an Antarctic killer whale, type B (*top right*); a leucystic (white furred, but not albino) Antarctic fur seal (*second row, left*); a leopard seal (*second row, right*); polar bears (*third row, left*); an Atlantic walrus (*bottom right*); and a sea otter (*bottom left*).

Two factors that have played a major role in shaping the anatomy, physiology, and behavior of dugongs and the various species of manatees are their unusual combination of an aquatic lifestyle and a herbivorous diet. Living in water requires certain attributes common to warm-blooded marine mammals, including sirenians; whales, dolphins, and porpoises (collectively called cetaceans); seals, sea lions, and walrus (called pinnipeds, or flipper-footed animals); polar bears; and sea otters.

At a glance, it is clear that these animals, with the exception of sea otters and polar bears, have adaptations that allow efficient and rapid movement through the water. For example, they tend to have a streamlined (i.e., fusiform, or spindle-shaped) body to reduce drag, powerful flukes for propulsion, reduced protuberances (such as large, thick limbs that could cause drag), and nostrils (or blowholes) that are directed upward to permit easy breathing while swimming forward.

A perhaps less obvious trait is that all of these animals attain a large, if not gigantic size, with several species of great whale surpassing 50 feet (almost 20 meters) in length and 50 tons (more than 45,000 kilograms)

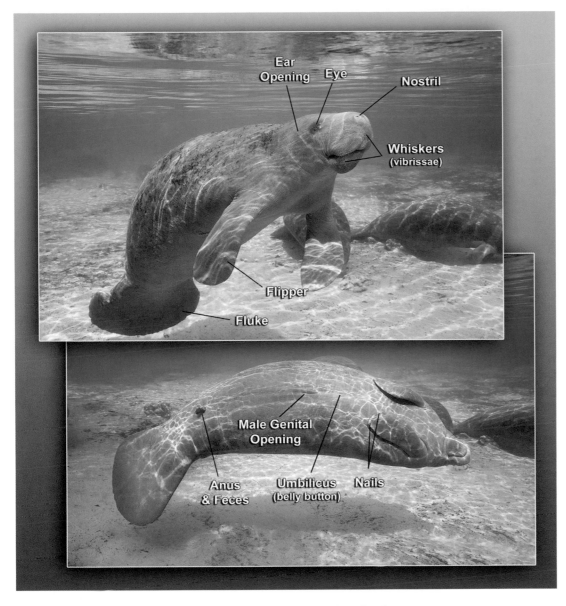

Ear Opening Eye Nostril

Whiskers (vibrissae)

Flipper

Fluke

Male Genital Opening

Anus & Feces

Umbilicus (belly button)

Nails

in weight. Although sirenians are not generally considered to be among the true giants of marine mammals, a recently exterminated relative of the manatees, called Steller's sea cow (*Hydrodamalis gigas*), may have exceeded 30 feet (nearly 10 meters) and 7 tons (6,300 kilograms). Of the living sirenians, the West Indian manatee is the largest species, with occasional individuals in the Florida subspecies approaching 12 feet (3.5 meters) and 3,600 pounds (1,620 kilograms). Even the smallest living sirenian, the Amazonian manatee, can measure almost 10 feet (2.8–3 meters) long and can tip the scales at nearly half a ton.

The link between large size and an aquatic lifestyle is this: bigger animals have a smaller surface-area-to-volume ratio than smaller ones, so large marine mammals have a reduced tendency to lose body heat across their skin, compared with littler mammals. This is extremely important, because water transfers heat approximately seventy times better than air, a fact that is apparent to humans who may be comfortable in a room where the temperature is 72°F but distinctly uncomfortable in a swimming pool at that same temperature. Simple physics means that a

The photograph depicts some anatomical features of a manatee. The animal in the bottom image is a male, as can be seen by the placement of his genital opening, which is rather far forward on his belly.

Florida manatees can reach an enormous size and girth. The largest and heaviest manatees tend to be females, and some individuals tip the scales at well over 3,000 pounds.

The density of hairs on the body of a manatee is greatest on the face and around the mouth. Sparse hairs are found all over the body, however. These hairs are sensitive to mechanical disturbances, such as water movements, and they appear to function for orientation and the coordination of movements, similar to the lateral line (pores on each side of the body, opening into sensory organs) in schooling fish.

large-bodied mammal is inherently better at retaining body heat than a small mammal.

Large size alone, however, may not be sufficient to keep a marine mammal from excessively losing body heat. In order to help retain their body heat, many marine mammals possess thick fur (sea otters, polar bears, fur seals) or thick blubber (whales, seals, walrus) as insulation. Manatees, however, have little external hair (not nearly enough for heat retention), and although they do have body blubber, it is arranged in layers that are separated by muscle and thus provides far less insulation than is the case for most other marine mammals of similar body size. Lacking these typical marine mammal defenses against cold, manatees have adapted by generally avoiding such conditions. They tend to occupy

tropical and subtropical waters, and, in locations such as Florida, where there is a possibility of relatively cold weather in winter, they migrate south or seek warm-water refuges, such as natural springs or, in modern times, areas where power plants discharge warm water.

As a rule, large mammals tend to have a lower weight-specific metabolic rate than small mammals. Thus a hamster needs to eat every day to survive, but a large mammal, such as a human or manatee, can fast for some time, another useful adaptation under certain circumstances. Florida manatees are an extreme case, exhibiting a remarkably reduced metabolic rate, measuring only about 15 percent of what might be expected from a mammal of their size.

Manatees also stay warm by producing extra body heat, due to the breakdown of cellulose (from plant material they consumed) by bacteria in their digestive systems. Florida manatees are more rotund than Antillean manatees, because the Florida subspecies has become adapted evolutionarily to occasional exposure to cold weather by having a larger gastrointestinal tract—in essence, a bigger furnace for heat production.

**Although Florida manatees have gigantic, ponderous bodies, their grace and speed underwater can hardly be exaggerated. They typically swim at speeds of 1.2–4.4 miles per hour (2–7 kilometers per hour), but they are capable of moving at 11.2–15.5 miles per hour (18–25 kilometers per hour) in short bursts, thanks to the incredible power of their fluke. In comparison, the fastest recorded swimming speed for a human is just above 5 miles per hour.**

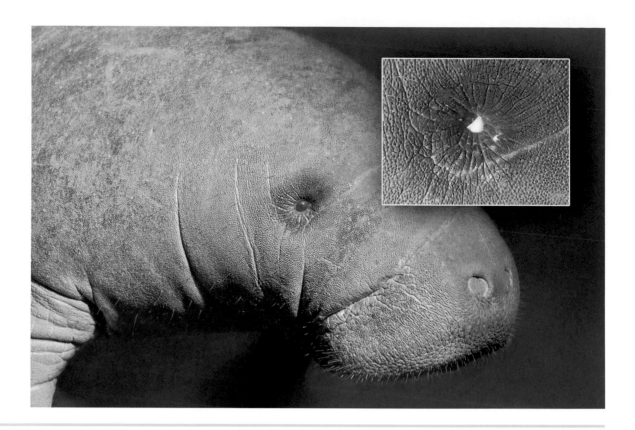

Manatees have a range of adaptations to allow them to live in the water. Among other things, they have eyes with protective nictitating membranes.

In addition, manatees (as well as other marine mammals and certain fish) possess counter-current heat exchangers, that is, anatomical arrangements of some small arteries and veins that allow them to efficiently prevent warmth (body heat) in their arterial blood from reaching the periphery of their bodies, where it would be more likely to diffuse to the environment.

Although many terrestrial mammals, such as humans, rely on vision to sense their surrounding environment and orient themselves, it is difficult to see for any distance in the aquatic world. Instead, manatees tend to rely on other senses. For communication over longer distances, manatees and many other aquatic species have come to rely on sounds (chirps, squeaks, and whistles, in the case of manatees). For orientation to their environment, manatees may use taste (a type of chemoreception), and, over short distances, they use nerves associated with the sparse hairs on their bodies to detect and respond to movements in the water.

In addition to adaptations associated with their sensory and communication abilities, manatees have specialized kidneys that help them achieve an internal water and salt balance (i.e., osmoregulate) in aquatic environments that range from seawater to freshwater and all salinities in between. Most aquatic species lack the capacity to adapt to and live in such a wide range of salinities (an ability termed euryhaline). The array of adaptations manatees possess facilitate living and moving freely in the water and allow them to keep warm during most cold spells.

West Indian manatees and their relatives, being large, require sub-

stantial food resources (in the form of aquatic plants) and thus need to live in habitats that support lush plant production. Inasmuch as manatees and dugongs often live in proximity to humans, their habitats (such as the Indian River Lagoon) can be significantly impacted by pollution, dredging, runoff, or other consequences of human activities that can lead to illness or death for resident species. Since fasting or poorly nourished manatees will not be able to produce much (or any) heat associated with the cellulose breakdown by bacteria in their gastrointestinal tract, a lack of forage has both nutritional consequences and serious temperature regulation ones, especially during cold spells in Florida.

An additional source of vulnerability associated with large size is that big species, such as manatees, tend to breed very slowly and recover relatively poorly from population reductions, because their life-history attributes, which Robert E. Ricklefs defined as "the set of adaptations of an organism that more or less directly influence life table values of age-specific survival and fecundity; hence reproductive rate, age at first reproduction, reproductive risk, etc." predispose them to maintaining

Not only are manatees more social than was once believed, but they are also quite curious and playful.

One of the most important threats facing manatees now and into the future is the loss or modification of their habitat. With Florida's human population projected to increase nearly 30 percent over the next fifteen years, finding ways to balance human needs and activities with a healthy environment and sustainability for wildlife is an enormous challenge for Floridians.

Female manatees generally have one calf per litter, and the bond between a mother and her calf is strong.

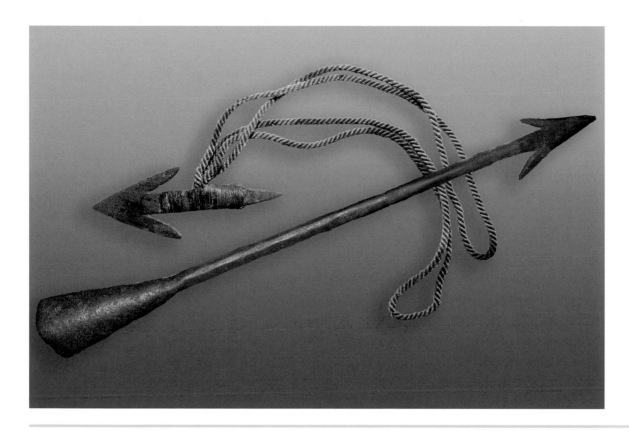

rather stable populations at or around the carrying capacity of a particular environment. Some other attributes that often, but not always, accompany large body size are a long lifespan, small litter size, delayed onset of sexual maturation, extended maternal or parental care, the ability to breed many times during their lifespan (called iteroparity), and relatively poor ability to colonize new habitats. To a greater or lesser degree, manatees and other sirenians possess these characteristics. This collection of life-history attributes presents a challenge for wildlife managers and conservationists if population levels become reduced through overharvest, incidental or accidental, habitat loss or degradation, or disease; for manatees and dugongs, it is almost biologically impossible to rapidly recover their population numbers.

Finally, it is important to note that human hunters tend to focus on large animals, which provide a substantial return for the work involved. For example, if it takes equal effort to harvest a 1,000 pound (~450 kilogram) manatee as it does to harvest 100 pounds (~45 kilograms) of fish, and if both resources have equal value as food, hunters in subsistence communities will typically opt for taking the larger individual, which has the greater nutritional and other immediate benefits. This has contributed to the overexploitation of many species of marine mammals by commercial hunters for hundreds of years. In addition, not only do manatees and other marine mammals represent a huge, tempting resource for hunters, but the fact that they are air-breathing mammals causes them to expose themselves periodically at the surface to breathe, adding to their vulnerability to hunters.

**To this day, manatees around the world are harvested for meat and other products. The harpoon shaft (*bottom*) is from Gambia, West Africa, and the detachable harpoon head (*top*) is from the Orinoco Delta, Venezuela.**

**Manatee lips can grasp plants and other objects; this capability defines the term "prehensile."**

For West Indian manatees and other sirenians, these sorts of vulnerabilities have contributed to their extermination from parts of their range and to a great reduction in their numbers in other areas. Nonetheless, manatees are highly resilient and adaptable in their behaviors and thus could be thought of as "aquatic coyotes" or "aquatic raccoons," capable, if given half a chance, of thriving in close proximity to people.

It seems clear that adapting to life in the water creates certain features that differ from the typical mammalian theme. Herbivory represents

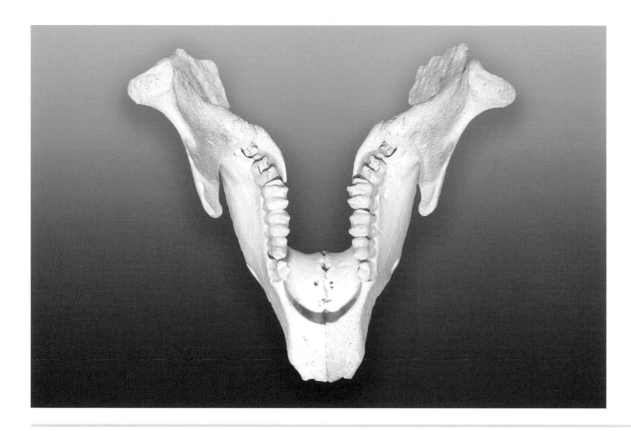

the other factor that has shaped manatee adaptations in unusual, if not unique ways. Although the word "unique" does get overused to represent special attributes of people or animals, it is safe to say that the collective digestive apparatus of manatees really *is* one of a kind.

A manatee's lips are prehensile, capable of grasping and manipulating plants and inserting them into the animal's mouth. To assist with this function, their lips have specialized hairs (called vibrissae) and oral bristles that are filled with an abundance of nerves. The vibrissae can even assist a feeding manatee to remove sediment from seagrasses or excavate rhizomes (underground plant stems). In some ways, the lips of manatees resemble and function similarly to the trunk of a related species, the elephant.

Once this plant material enters the mouth, molar-like teeth grind it down, but not without cost. In the process of chewing their very gritty food, manatees wear down their teeth. To compensate, manatees have the ability to replace teeth for virtually their entire lives. This is accomplished by having the six to eight teeth in each quadrant of the jaw migrate forward slowly, as if on a conveyor belt. As the front teeth get increasingly worn down, they eventually lose their usefulness and fall out of the front of the mouth. New teeth, with sharper grinding ridges, are continuously produced at the back of each tooth row.

As food moves down into the manatee's gastrointestinal tract, the relative proportions of that tract provide an enormous capacity for the breakdown of ingested materials and the absorption of nutrients. Although a manatee's stomach is not disproportionately big, its large in-

**This photograph shows a set of teeth in the lower jaw of a manatee. As worn-down teeth eventually fall out, there are always new, sharper teeth emerging from the back of each tooth row.**

Although algae are not usually a main component of the diet of a Florida manatee, this animal is clearly showing that almost anything green is acceptable to eat. Adult manatees eating hydrilla (a freshwater exotic plant) have been estimated to consume the equivalent of around 7 percent of their body weight per day. For a 1.5-ton manatee, this would equate to approximately 210 pounds (462 kilograms) of plant material per day.

testine, when filled with digested material, can weigh up to 330 pounds (nearly 150 kilograms) and measure as much as 20 meters (60+ feet) long. The proportions of the manatee's tract are generally similar to what is found in some other non-ruminating herbivorous mammals, such as horses and elephants.* Other non-ruminating species (also called hindgut digesters, because the digestion of cellulose occurs predominantly in the large intestine, or hindgut), however, lack a distinctive manatee feature: a specialized structure, called a cardiac gland, protruding from the side of the stomach in the general direction of the heart. The cardiac gland is found in fewer than a dozen other species, and its function is to sequester digestive enzymes and acid-producing cells away from the very abrasive food those species consume. Manatees also have some unusual cellular arrangements in the lining of the remainder of their gastrointestinal tract.

The breakdown of cellulose by bacteria is obviously most useful to a herbivore, because the process makes nutrients available for the host. Hindgut digesters tend to eat almost constantly, so they require a great deal of food; as a trade-off, they can subsist on low-quality forage. Ruminants have much a smaller digestive system and pass food through the system rather slowly; therefore, they need higher-quality forage to survive. Competitively, hindgut digesters can thrive on a diet that would nutritionally compromise a ruminant.

Heat and methane gas are byproducts of the breakdown of cellulose

*A cow is a typical ruminant. Such species have four stomach compartments where fermentation of ingested plant material occurs. Cows have a modest-sized large intestine.

Manatees and dugongs are considered to be hindgut digesters. Such species consume plants, but most of the digestion of the material they eat and the absorption of nutrients occur in the hindgut (i.e., the large intestine and its cecum, or intestinal pouch). This stands in contrast to what happens in cows and sheep, which are classified as ruminants, because most of their plant digestion occurs in a massive, multichambered stomach. The animals shown here include some other hindgut digesters: a black rhinoceros *(top left)*, a plains zebra *(top right)*, African savannah elephants *(bottom left)*, and a hippopotamus *(bottom right)*.

At some warm-water refuges, the seafloor is barren. In certain locations, this can result from the volumes of warm water produced, in combination with already-tepid water during the summer.

Manatees have a surprising capacity to adjust their buoyancy to remain at the water's surface or to rest midway down or on the bottom. The unusual structure and thickness of the diaphragm, the presence of long lungs that extend almost the entire length of the body cavity, and gas production associated with digestion all contribute to their excellent buoyancy regulation.

from the plants that a manatee consumes. The former is important for the survival of Florida manatees, since feeding manatees are much better able to stay warm during cold weather in winter than are fasting individuals, but some of the warm-water refuges occupied by these animals in winter only give them limited access to submerged plants.

The production of methane, which can be lost through flatulence, actually helps save energy for manatees by increasing their buoyancy. It has even been suggested that the structure and position of a manatee's diaphragm, coupled with the release of methane through flatulence, helps manatees maintain their buoyancy and ascend and descend in the water column (a conceptualized column in a body of water, extending from the water surface to the bottom sediments) with no apparent effort. An additional outcome of the manatee digestive processes is the production of abundant feces. Manatee fecal material fertilizes the seagrass areas that the animals frequent, making those locales more productive.

Being an aquatic herbivore has shaped not only certain anatomical structures for these animals, but it has also influenced manatee behavior

and habitat use. For example, submerged aquatic plants necessarily grow in shallow water, where light can penetrate well and allow photosynthesis. For that reason, coupled with the fact that, as hindgut digesters, manatees forage for many hours per day, the animals occupy inshore, shallow habitats and have not evolved any remarkable diving capabilities, relative to many other marine mammals.

Altogether, it is clear that over tens of millions of years of natural selection in an unusual ecological niche for a mammal, manatees have been shaped into a creature with striking adaptations and proportions. Those adaptations entail an outstanding array of morphological (related to form and structure), physiological (related to physical function), behavioral, and sensory specializations. The resulting hodge-podge is therefore not inappropriately called "a bewhiskered blimp of an animal."

This chubby manatee displays some of the distinctive features that contribute to the iconic perception of the species: a bewhiskered face, small eyes, paddle-like flippers with nails, a rotund body, and an apparent smile.

Many people think that manatees do nothing but float around and eat. On the contrary, they interact regularly with one another and with various objects in the water.

# 3

# *Evolution*

As recently as the mid-eighteenth century, five species of sirenians existed: three species of manatees (Amazonian, West Indian, and West African manatees) and two of dugongids (dugongs and Steller's sea cows). In 1741, shipwrecked Russian hunters and explorers discovered Steller's sea cows in the Commander Islands (the westernmost extension of the Aleutian Islands). But available records indicate that by 1768, a remarkable span of just twenty-seven years, sea cows had been exterminated by fur hunters (primarily from Russia), so only four sirenian species remain today. The distribution of the four living species includes tropical and subtropical waters of five continents: Asia, Africa, Australia, North America, and South America. In contrast, the distribution of the gigantic Steller's sea cows was confined to the frigid waters of the North Pacific Ocean.

The living sirenians are united by several morphological attributes. All possess fusiform bodies that lack pelvic (hind) appendages entirely and have reduced pectoral (front) limbs. They all use a horizontally expanded fluke—rounded in manatees and split in dugongs (as it is in dolphins and whales)—for propulsion. Their bones are pachyostotic (thick and swollen), osteosclerotic (very hard and solid), and quite heavy. The sirenians lack an externally distinct neck, have a small brain (relative to their body

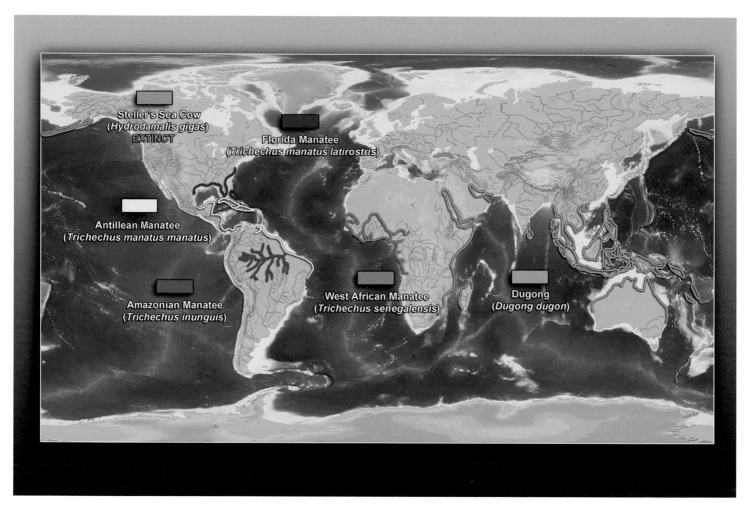

West Indian (Florida and Antillean) manatees have a wide range, outlined here in red and yellow. It is next to nothing, however, when compared with the range of dugongs, which spans the waters of thirty confirmed states and territories, with an additional half dozen states also likely to have dugongs. The range of West African manatees is small and noncontinuous, and the status of this species is probably the most precarious among the living sirenians.

size) and cranial cavity, and possess specialized dentition (teeth). The sirenians have thick hides and sparse hairs over their bodies.

All of the living sirenians reach a very large size, with some Florida manatees (the largest extant taxon) occasionally approaching almost 12 feet (3.5 meters) in length and 3,600 pounds (1,620 kilograms) in weight (roughly as heavy as a small car or twenty large adult humans weighing an average of 180 pounds each). West African manatees generally resemble their West Indian relatives, but reliable reports of maximum weight and length for the former do not exist. The smallest manatee is the Amazonian manatee, which reaches a maximum length of around 9 feet (2.8 meters) and a maximum weight of about 1,050 pounds (a little more than 450 kilograms). Dugongs typically reach a length of 9 feet (approximately 2.7 meters), but individuals reliably measured at 11 feet (3.3 meters) long have been reported. Dugongs are more slender (thus appearing more "athletic") than manatees, and 10-foot-long adults weigh "only" about 880 pounds (400 kilograms).

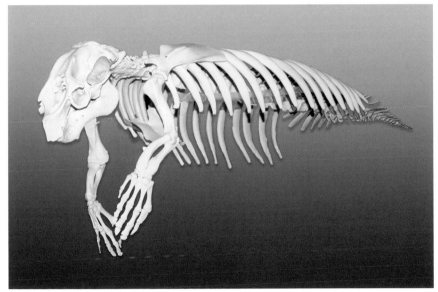

A fluke is not exactly the same thing as a tail: the former lacks bones, whereas the latter has them. The most aquatic marine mammals—namely, whales, dolphins, porpoises, and sirenians—possess either a single fluke (like the manatee) or split flukes (like a dolphin or dugong).

The skeleton of a manatee is noteworthy, compared with our own, for at least a couple of obvious reasons. First, a manatee lacks leg bones. Second, manatee bones are extremely thick and swollen and are quite heavy. Although it cannot be discerned from this photograph, the bones are also solid and hard.

The living sirenians and their immediate ancestors have been shaped by natural selection for the past 60 million years, ever since some terrestrial mammals (called paenungulates) reinvaded aquatic habitats and became adapted to a very unusual niche as herbivorous, totally aquatic mammals. Some of the manatee's physical attributes, however, are also associated with inherited traits from its terrestrial ancestors.

The Afrotheria is an odd grouping of animals made on the basis of their presumed evolutionary history, consisting of a common ancestor and all of its descendants; such groupings are called clades. The Afrotheria includes the sirenians, as well as such morphologically diverse

The Afrotherian clade includes an unusual assemblage of related animals. Besides sirenians, clade members include species such as the aardvark (*top*), African savannah elephant (*bottom left*), and tree hyrax (*bottom right*).

mammals as elephants (order Proboscidea), aardvarks (order Tubulodentata), hyraxes (order Hyracoidea), tenrecs (order Afrosoricida, family Tenrecidae), golden moles (order Afrosoricida, family Chrysochloridae), and elephant shrews (order Macroscelidia). Although the members of the Afrotheria do possess some common morphological features, the use of modern tools in molecular biology, genetics, and biochemistry have provided even stronger support for and cemented the affinities among this bizarre group of mammals.

Most members of the Afrotheria are small and clearly unlike manatees, but elephants and manatees have a number of obvious anatomical and life-history similarities, including but not limited to the following: large body size, long life span, delayed onset of sexual maturity, small litter size, extended maternal care of their young, thick skin, axillary (located near the armpit) mammary glands, abdominal testes, and a lengthy gastrointestinal tract, as well as some osteological (bone) characteristics, aspects of the placenta, tooth structure, and mechanisms for tooth replacement.

Oddly enough, sirenians are not closely related to the other marine mammals, although they certainly have some morphological characteristics in common. In contrast to manatees and dugongs, however, cetaceans (whales, dolphins, and porpoises) are related to artiodactyls (even-toed ungulates, such as hippos and cows), pinnipeds (seals, sea lions, fur seals, and walrus), polar bears, and sea otters are derived from carnivorous ancestors, such as bears and mustelids (e.g., weasels and minks). Superficially, sirenians appear most like the walrus, but there are several differences between them, such as walrus being partial to mollusks for food.

The sirenians and a related group, the desmostylians (all of which are extinct), broke away from their terrestrial relatives around 60 million years ago, according to some sources. The oldest definitively sirenian fossils are roughly 50 million years old, from the Eocene epoch. It has been suggested that the evolution of sirenians from their fully terrestrial ancestors may have occurred in marshy areas containing abundant floating or emergent plants, on which the animals foraged. Although

Until the latter part of the nineteenth century, many zoologists considered manatees to be odd, subtropical forms of the walrus. One look at their similar faces, gigantic size, and stout bodies shows why. The manatee's close relative, the dugong, actually has tusks.

Two Florida manatees cruise over a meadow of turtle grass, a common seagrass in Florida and in manatee diets.

the oldest sirenian fossils were found within deposits in Jamaica, it is believed that this taxonomic group arose in Eurasia and/or Africa.

Sirenian diversity is low at present, with only four living species. The fossil record, however, demonstrates that sirenians were far more diverse in the past, especially in the North Pacific Ocean, where a number of dugong-like species existed, and in the Caribbean, where four families and many species of ancestral manatees and dugongs lived and competed. Natural selection led to the Caribbean manatees eventually (i.e., around 5.3 million years ago) outcompeting Caribbean dugongs and other early (and now-extinct) types of sirenians; thus the manatees became and remain established in that region.

The earliest animal that had a truly manatee-like appearance is named *Potamosiren* and dates from the Miocene epoch, approximately 15 million year ago. Although *Potamosiren* generally resembled modern manatees, they lacked the extra teeth and the tooth replacement mechanism that characterize today's manatees. After the middle Miocene, the trichechids (manatees and manatee-like animals) developed this modern set of dental features (dentition).

In the wider Caribbean area, the evolution of a range of seagrass species coincided with the expanded diversity of the sirenians in the Oligocene and Miocene epochs. It seems likely that the manatees' type of lifelong tooth replacements helped them outcompete the resident dugongs, as the level of ingested silt took a toll on the teeth and the consequent foraging ability of the latter, thereby opening the door for the predominance of various ancestral manatees.

Approximately 1–2 million years ago, a species of manatee emerged from the wider Caribbean region as the dominant marine or coastal sirenian in that part of the world (Amazonian manatees are mainly found in freshwater habitats of the Amazon and Orinoco river basins and deltas). Today, scientists have confirmed that this ancient species actually represents the modern species called the West Indian manatee, and over time it came to occupy much of the present range of this species—basically in coastal and riverine waters from the southeastern United States to northeastern Brazil. For comparison, our own species (*Homo sapiens*) has existed for a scant 0.2 million years. This temporal relationship gets to the heart of bizarre anticonservation arguments made by some opponents of Florida manatees about whether the species is native to Florida, with the corollary that, if it is not, wildlife managers and others should consider manatees to be exotic and not attempt to protect and sustain the species. The response is that *humans* are a nonnative species in Florida and the Caribbean, and people should therefore be careful about wishing to eradicate the state's "newcomers."

**Silhouetted against the rays of the sun, a large manatee swims tranquilly through turquoise waters in Florida.**

Manatees are very curious, and swimmers and divers
are often surprised when one suddenly shows up face to
face. It is an experience to be treasured, but not abused
by chasing or otherwise harassing the animal.

# 4

# *The Stuff of Myths and Legends*

Given the nearshore and riverine habitats occupied by manatees, along with the edible and other products they provided for coastal communities in the prehistoric and more recent past, it is not surprising that manatees and their harvest became the stuff of myths and legends. Some of the traditional, long-standing perspectives remain, especially in subsistence communities, but in recent years new perceptions—and mispercep-tions—have arisen. Although it has been common for humans to focus on ways in which manatees can be used for or be useful to human en-terprises, a new ethic has started to arise that highlights ways in which people and manatees can coexist, to the benefit of both. This chapter explores a range of examples to illustrate human values and perceptions toward Florida manatees and other sirenians.

It is clear that for most of the history of manatee and dugong inter-actions with people, the former two species were viewed primarily as sources of meat and other products for the latter. Subsistence-based societies generally maintain their cultural values through an oral, rather than a written tradition, so it is not known exactly how long people have harvested manatees and other sirenians. It is revealing, however, to consider that the biblical Ark of the Covenant has been suggested to

In addition to the up-and-down motion of its fluke, the flippers of a manatee are unexpectedly flexible.

have been bound in a covering of dugong hide. In Florida, the harvest of manatees has been documented from digs that have occurred since the Paleo-Indian period (8500–6000 BC), coinciding with the first known occupation of the state by aboriginal indians. Dugong hunting in the Torres Strait (between Australia and Papua New Guinea) has been documented to have occurred for some 4,000 years, and one site in the strait (on Mabuiag Island) contains the remains of an astounding 10,000–11,000 dugongs, harvested over a 300-year period (dating from around AD 1600 to AD 1900). Clearly, humans and sirenians have a long relationship, from which the former derived substantial benefits.

Despite some of these remarkable harvest quantities, the impacts of subsistence-level hunting usually reflect, to some extent, the size of the human population when the harvests occurred. At the time Europeans arrived in Florida (about AD 1500), around 25,000 aboriginal indians commonly are estimated to have been in the state. The taking of a large and powerful manatee represented quite a task, and it is uncertain what sort of impact these indians had on the population of manatees at that

time. In management and conservation, however, it is important to maintain a perspective of scale: 25,000 men, women, and children scattered throughout the commodious state of Florida represents a low-density human population, approximately one-third the number of the people present at a Sunday Tampa Bay Buccaneers football game. Thus most of the truly detrimental impacts of hunting probably occurred after the growth of human populations in areas occupied by manatees, the development of powerful weapons, and the appearance of commercial markets for certain products.

This observation was incorporated into some ecological and conservation principles originally articulated by the team of Paul Ehrlich and John Holdren, whose formula I = PAT attracted considerable attention, starting in the 1970s. The equation indicates that the environmental impacts (I) of humans are proportional to human population size (P), affluence (A), and technology (T). Although criticized for being overly simplistic, the equation nonetheless provides a dire reminder that, all else being equal, the burgeoning, technologically advanced human population on Earth today has a greater capacity for environmental damage and species extirpations than was ever the case in the past.

The most enduring myth about manatees is that they represent the Sirens described in Homer's *Odyssey* and other ancient writings. The Sirens are better known today as mermaids. It is hard to conceive that anyone ever thought that a manatee resembled a beautiful and alluring sea creature. It has been suggested that a manatee with seaweed on its head might somehow be mistaken for a mermaid, and the presence of

A 1909 postcard from Cameroon shows people with a freshly harvested manatee. Even as late as the end of the twentieth century, manatees were occasionally eaten by Floridians.

For the early mariners who compared manatees to the mermaids (Sirens) of mythology, the profile shown here could have effectively squelched the legend. Though manatees are beautiful in their own way, their resemblance to mermaids is, well, exaggerated.

axillary mammary glands in both mermaids and sirenians has also been offered as a reason to relate the two. Perhaps the best explanation is that sailors who perceived a resemblance between a mermaid and a manatee had simply been at sea far too long. In any event, the alleged relationship between mythical sirens and manatees and dugongs led early zoologists to classify the latter animals as the order Sirenia.

In the mythology of some cultures, however, sirenians are relatively frequently said to have arisen when people (often a young woman or, less often, a young man) drowned at sea. As a further link, it is alleged to this day that hunters and fishermen in some subsistence communities have intercourse with harvested female dugongs.

More commonly, there is no allusion to manatee beauty or sensuality, and the animals were and are currently harvested simply to provide products that can be consumed, used in other ways (e.g., the skin for leather, the fat for cooking or medicine, the thick and heavy bones for carvings or weapons) by the hunters, or sold or bartered. Historical records and

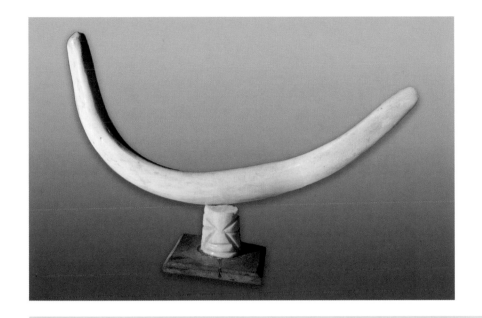

evidence from archeological digs indicate that West Indian manatees were once sufficiently abundant to provide an important supplement to, if not a staple in the diet of people in coastal communities around the wider Caribbean region. The harvest of West Indian manatees appears to have reached its peak during the colonial era (AD 1500–1700+), when both Amazonian and West Indian manatees in Brazil and other countries were taken in large (cumulatively, perhaps in the *hundreds of thousands*) but undocumented numbers by settlers, buccaneers, armies, and commercial ventures for meat and other products.

In Florida, the harvest of manatees by settlers, aboriginal indians, and others (e.g., commercial hunters for bird plumage and other marketable products) through the nineteenth century led to such noticeable reductions in manatee numbers that the state passed a law in 1893 to protect the species. Since this law was largely unenforced, it did little to stem the taking of manatees, and by 1973, the West Indian manatee was included for protection under the US Endangered Species Act, as well as the US Marine Mammal Protection Act of 1972. As recently as the 1980s, however, butchered carcasses of manatees were still reported (albeit rarely) in Florida.

Among the manatees, the species that was most heavily exploited for commercial reasons was the Amazonian manatee; its documented take lasted for centuries. In 1660, Father António Vieira recorded that twenty Dutch ships per year were provisioned with manatee meat (in this case, from either Amazonian or Antillean manatees) by the Nheengaiba

Manatee ribs are extremely heavy and dense, and, unlike those of humans, lack a marrow cavity. These qualities have historically made manatee ribs ideal for carvings. In this photograph, an intact curved rib is attached to a base of carved ivory or bone, with the latter being an ancient art form called scrimshaw.

Sometimes the clarity of Florida's coastal waters is amazing, permitting a view of marine organisms large and small. Manatees are among the more charismatic of the coastal marine species, and effective conservation of manatees will automatically conserve other, less appreciated members of the marine community as well.

Indians near Cabo do Norte, Brazil. The harvest continued unabated through the eighteenth century; in just two years in the mid-1780s, the Pesqueiro Real (Royal Fishery) da Villa Franca produced 3,873 *arrobas* (128,000 pounds, or approximately 58,000 kilograms) of dried and salted Amazonian manatee meat and 1,613 *potes* (70,000–106,000 pounds, or about 32,000–48,000 kilograms) of lard. These values suggest a harvest of around 1,500 manatees in two years. From the mid-1800s until the 1930s, the harvest increased, primarily for *mixira*, fried manatee meat packed in its own fat. The *mixira* was generally consumed locally (i.e., in Amazonia). It has been estimated that several thousand manatees were harvested annually for their meat and fat. The most recent documented commercial market for Amazonian manatees took place between 1935 and 1954, when as many as 7,000 manatees were harvested per year. In this case, the product that was most highly sought was their durable, tough hide, which was used for machine belts, hoses, and gaskets.

Today, hunting of West Indian manatees and other sirenians continues, although the generally low population density of the animals limits

the number of animals taken. In most cases, the harvest of sirenians is illegal, although some subsistence communities are allowed by law to maintain a traditional hunt. In certain South and Central American countries, the currently illegal harvest appears to take place along traditional lines. For example, on a recent trip to the Bocas del Toro region of northeastern Panama, I was fortunate to have spent time with scientists and naturalists in a dugout canoe, patrolling rivers known to harbor manatees. Bocas del Toro juxtaposes wonderful natural flora and fauna with gigantic commercial banana plantations. Here and there along the waterways, we spotted platforms built in the mangroves, and I was told that manatee hunters would traditionally tie bunches of green bananas just above the surface of the water below such platforms. When a manatee was lured to feed on the fruit, it exposed its head, neck, and back and could be easily dispatched with a harpoon or gun. Scientists in Panama who wish to capture manatees for tagging, health assessments, or other forms of research use the same tried and true approach (without killing the manatee, of course): when a manatee starts to eat the bananas, a net is quietly pulled in place to encircle the unwary animal.

In Florida, scientists sometimes use a variation on this theme to capture manatees for study. Taking advantage of the fact that Florida manatees like to drink freshwater from hoses at marinas and elsewhere, researchers who hold appropriate federal permits suspend a hose above the surface of the water, wait for an animal to start drinking, and slowly pull a net in place behind it, cutting off any escape.

Sirenians occupy the waters of a number of developing countries

Manatees breathe every one to eight minutes, depending, in part, on their activity level. Sleeping manatees awaken to breathe.

Bocas del Toro, located in northeastern Panama, has a combination of beautiful beaches, jungles, and massive banana plantations. In remote locations in South and Central America, manatees, sea turtles, and other species may be poached for food and other products.

that rank very low on the Human Development Index (an average of measurements of life expectancy, education, and the standard of living in a country), and the harvest of single one of these unaggressive, slow, huge animals in a poverty-stricken community can nourish that community and dramatically increase the annual income of its members. This is true in places where the taking of sirenians is legal, as well as those where it is not. In such circumstances, the presence of protective

In many parts of the world today, people cherish any interactions with manatees and are staunchly protective of the animals. Historically, however, manatees and dugongs have been appreciated more for the products they provide, including their meat. Hunting and poaching remain significant threats to manatee populations in many parts of the world.

legislation, by itself, does not guarantee that manatees and dugongs will be left alone.

It is important to realize that manatees and dugongs have not lost their cultural importance in some subsistence societies. Thus in some communities where hunting is perfectly legal, such as in the Torres Strait Islands (between Australia and Papua New Guinea) and northern Australia, enhanced social status is accorded to successful hunters of manatees and dugongs. In most locations occupied by sirenians, however, the development of protective laws makes it imprudent to advertise which community members are active hunters. In such countries, legal hunting with quotas is often replaced by opportunistic and unregulated poaching, which can seriously impact not just small populations of manatees, but can even substantially reduce relatively large ones.

Although manatees have generally been viewed by people as sources of products or providers of enhanced social status for successful hunters, that has not always been the case. In Cameroon, West Africa, for

As charismatic animals, manatees attract considerable
attention, which is linked with their conservation.

example, manatees are considered by some native groups to be extremely
dangerous and are to be avoided at all cost. On the other hand, in certain
countries manatees have recently come to be perceived as important
allies for various human stakeholders. The concept that manatees can
be useful to people without harvesting or consuming these animals is
very beneficial for conservation efforts. For example, manatees and dugongs are considered to be flagship species, where effective efforts will
simultaneously conserve vital habitat for them and for other species of
ecological, aesthetic, or economic value. Whereas it may be difficult to
engender popular support for the conservation of species of snails or
seagrasses, well-conceived and comprehensive strategies for the conservation of charismatic manatees will automatically and consequently
have broad benefits for a range of species, including humans.

In a somewhat similar fashion, grazing by manatees can make the
species extremely useful to a number of human stakeholders. In some
locations (most notably Guiana), manatees are used to keep canals and
other waterways clear of aquatic weed growth, cost free and without

Manatees often have algae and small creatures living on their skin. Thus it is not uncommon to see fish, such as the school of striped mullets shown here, picking at a manatee's skin. It is a mutually beneficial relationship (symbiosis) in which the fish get a bite to eat and manatees rid themselves of freeloaders.

The fish that attend to manatees are sometimes quick and brightly colored. This photograph is reminiscent of a cleaning station on a coral reef, where neon gobies and other darting "jewels" rid host fish of parasites.

A manatee "hugging" a diver. At places such as Crystal River, diving with manatees provides an opportunity for people and manatees to interact closely. The experience is completely legal, as long as people let the manatees approach them, rather than chasing the animals or otherwise forcing an interaction. Appropriate human behavior around manatees and other Florida marine mammals is discussed in *Dolphins, Whales, and Manatees of Florida: A Guide to Sharing Their World*, by John E. Reynolds III and Randall S. Wells.

using toxic herbicides. In a broader example, manatees and dugongs are considered to be cultivation grazers, which means that by cropping aquatic vegetation, they actually promote its growth and diversity, which in turn supports what are known as nursery areas for fish and other species of ecological, commercial, and recreational importance.

Manatees are also useful to people who simply appreciate seeing them in the wild, but people who unduly abuse the privilege of being in the water with manatees may be in violation of federal laws and thus subject to stiff penalties. Ecotourism can also have some unforeseen issues. For example, there are ongoing discussions about regulating access to manatees around warm springs in Crystal River, Florida. Tens of thousands of people each year seek the experience of swimming or diving with the manatees there, and that number alone may stress the animals, since too many admirers can displace the manatees from the warm-water resource they need to survive during the winter. Responsible divers will do all that they can to watch quietly and let the animals do as they wish.

As a general premise, however, appropriately conceived and regulated ecotourism can enhance peoples' enjoyment of and appreciation for having manatees nearby—a fledgling step, but a step nonetheless in the emergence of a conservation ethic. In an early attempt to engender such an ethic, Daryl Domning eloquently described why manatees are valuable to people in an essay entitled "Why save the manatee?" Among others, he listed aesthetic, genetic, ecological, and economic reasons.

Even though most historical interactions between people and manatees were detrimental to the latter, human activities can lead to the protection of large numbers of Florida manatees, while also providing an opportunity for interested people to see the animals up close. The best example of this relationship involves the creation of warm-water refuges in the discharge zones of some coastal power plants. Several hundred animals, ranging from adults to tiny calves, seek such areas when winter's cold weather arrives, and the availability of a predictable source of warmth can save many manatees from death or illness. In addition, having so many animals in a place where they can be predictably viewed and appreciated by the public creates a great opportunity to provide education about manatees and their conservation. Locations such as the Florida Power & Light Company's power plants at Riviera Beach and Ft. Myers, Florida, and the Tampa Electric Company's power plant at Apollo Beach (near Tampa) provide interested people with a chance to see hundreds of manatees just a few feet away.

Approximately 4 million years ago, the genus *Homo* evolved, and our own species, *Homo sapiens*, arose a scant 0.2 million years ago. Primitive

The Tampa Electric Company's power plant in Apollo Beach, Florida, is among the industrial sites that produce warm water that attracts manatees in winter. A large number of manatees (*inset*) are gathered in the water by a Florida Power & Light Company power plant located in Riviera Beach, Florida.

Like baby animals everywhere, manatee calves enjoy being close to their mothers. Sometimes, this even results in a free ride on her fluke.

people hunted sirenians for food and other products, and subsistence communities continue to do so to this day. It appears that no sirenian species was ever wiped out by hunters in subsistence-level communities, although local stocks are likely to have been extirpated in this manner. At the species level, the only sirenian known to have been exterminated by humans was the Steller's sea cow.

Thus as some past practices still persist, and as technology has evolved, directed hunting has the potential to entirely wipe out or contribute to the extermination of manatee population stocks, or even entire species. Of all the extant sirenian taxa, the only one for which hunting does *not* constitute a major threat is the Florida manatee. In the modern era, however, it seems likely that the greater danger to most manatee and dugong populations arises from the destruction or significant modification of the habitat they need to survive.

It is best for manatees if people observe them or allow the animals to approach, rather than initiate contact with these marine mammals in the wild.

A large manatee can eat a great deal—up to 200 pounds or more of vegetation each day. One of the benefits of being so big, though, is that these animals are able to fast. When necessary (e.g., while occupying a warm-water refuge during prolonged cold weather), a manatee can go without food for several days.

# 5

# *Behavior and Ecology*

Imagine your life being dominated by needing to consume 8–10 percent of your body weight every day, just to maintain yourself. If you weigh 200 pounds, try just on one day to eat 20 pounds of food—and then imagine eating that amount every day. Moreover, think of the time involved: roughly a full workday, doing nothing but eating. Welcome to the world of the Florida manatee.

Manatees are hindgut digesters, like horses and elephants. They can subsist on low-quality forage, unlike more-fastidious ruminants, such as cows. But manatees must eat steadily throughout the day, extracting every possible ounce of nutrition from every bite.

As a result of the manatees' evolutionary past, taxonomic ancestry, and distinctive morphology, their daily life is dominated by food and feeding. Luckily for them, life is a smorgasbord! West Indian manatees (both subspecies combined) have been documented or inferred to consume nearly 100 genera (a scientific classification that includes a group of closely related species)—and thus a much larger number of species—of freshwater and marine plants. These plants include algae, seagrasses, mangrove leaves and seedlings, and submerged, floating, and emergent (growing in or along the edge of the water, but with most parts in the

Hindgut digesters like manatees can subsist on low-quality vegetation (such as algae), if necessary. Although manatees need to consume a great deal of low-quality food, the fact that they are hindgut digesters allows them to occupy and forage in a wide range of habitats.

air) freshwater plants. Manatees consume both native plant species and exotic, introduced ones; their diet also includes some animals (e.g., tiny organisms living on plant leaves and even [very rarely] fish). The West Indian manatees' distribution across waters with a wide range of salinities contributes to the breadth of their diet; so does the modest degree to which the skull tips downward toward the nose and mouth, which allows these animals to forage at any level in the water column, at its interface with sediments, or even on land. In contrast, dugongs have a considerably downturned snout, necessarily making them bottom feeders. Manatee foraging occurs most commonly in waters ranging from 3 to 16 feet (1 to 5 meters) deep, not because manatees are that limited in their diving ability, but because submerged plant growth is more luxuriant in shallow areas, where there is excellent light penetration for photosynthesis to occur.

Although Florida and Antillean manatees consume a wide range of plant species, their food preferences vary with habitat and the consequent availability of edible materials. Manatees are rather selective in their foraging, a fact that strikes me when I conduct aerial surveys to count manatees and assess their habitat use patterns. For example, in Charlotte Harbor and Pine Island Sound in southwestern Florida, the shallow waterways provide excellent habitat for manatees, with abundant seagrass beds; several rivers with freshwater to drink; nearby locations with warm water in winter; good water quality; and quiet canals and small, shallow, bay-like areas. The seagrass beds cover a large percent-

Florida manatees consume a wide range of plants, including algae (*top left*); seagrasses (*top right*); the exotic, freshwater plant milfoil (*middle*); and seedlings of red mangroves (*bottom*).

age of the bottom, but the manatees I see tend to focus most of their foraging activities in just a few spots, such as in San Carlos Bay and off the north end of Pine Island. Given the vastness of the resource, why do the animals restrict where they feed?

Seagrasses and other marine plants are not very nutritious; I expect that a marine ruminant (if one existed) would starve, because ruminants need high-quality forage to survive. But even hindgut digesters such as manatees prefer better-quality food, if it is available. Thus manatees and dugongs select plants that are relatively high in nutritional value, including the amount of nitrogen the vegetation contains. New growth in seagrass genera such as *Halodule*, *Thalassia*, and *Syringodium* suits manatees very well, and the latter seagrass is so often sought as forage by the animals that its common name is manatee grass.

As manatees munch on seagrasses and other plants, the cropping process actually stimulates growth and productivity in the plants being consumed. People who mow their lawns are aware of this phenomenon:

Here, a foraging manatee comes quietly to the surface for a quick breath of air. For huge animals, manatees can create almost no signs that they are present. Although foraging manatees are sometimes difficult to see, if you put your head underwater, you will often hear them chewing.

when a lawn is mowed regularly, it grows quickly and the blades of grass tend to be relatively soft, but an untended lawn stops growing when the blades reach a certain height, and the blades themselves become tougher and more fibrous.

When manatees in Charlotte Harbor and Pine Island Sound return to the same seagrass beds again and again, it is because their regular cropping of the plants in those spots has made the grass blades more nutritious and easier to eat. Therefore, manatee foraging actually increases the productivity of seagrass beds and other vegetation in Florida waters. More-productive seagrass beds, in turn, provide better habitats for juvenile fish and other organisms that have a high aesthetic value or are important for recreational and commercial fishing. Due to water-quality problems and algal blooms in Brevard County in recent years, a seagrass die-off had taken place, with the loss of nearly 50,000 acres in 2011 alone. The estimated value of those seagrasses for ecosystem services (broadly defined as the benefits people derive from ecosystems) per acre per year to local economies was an astounding $5,000–$10,000, so the 2011 seagrass die-off created a potential local loss of $235 million to $470 million *for each year in which the reduction in seagrasses persisted.* Obviously, any factors that decrease seagrass coverage and productivity have major negative implications for coastal economies, whereas any factors that promote such productivity have equally significant positive ones.

The presence of foraging manatees, therefore, has a direct and positive impact on the economic value of coastal ecosystems for humans.

Manatees and dugongs are not the only grazers on marine plants. Among the worldwide species of marine grazers are the Galápagos marine iguana (*top left*), North American black brant (*top right*), South American black-necked swan (*bottom left*), and green sea turtle (*bottom right*).

This concept surprises many individuals, who assume that a growing manatee population will eliminate seagrasses, promote turbidity, reduce fisheries resources, and otherwise diminish ecosystem quality and the benefits humans obtain from it. That is simply not the case.

Nonetheless, there are circumstances in which large numbers of manatees can temporarily overgraze local vegetation. For example, when hundreds of manatees gather together at a warm-water refuge during severe or prolonged cold weather in winter (a grouping that is called an aggregation), those animals disperse to feed when the weather moderates. Typically there is not much submerged vegetation in the immediate vicinity of Florida's power plants, because the water discharged in summer, when ambient water temperatures in inshore waters can exceed 100°F, is sufficiently hot to impair localized plant growth. In winter when the animals leave the warm-water refuge to forage, they typically seek nearby vegetation, which can be grazed down to the sediment. If the plant roots and rhizomes remain, however, regrowth occurs when the weather warms up and the manatees depart.

Outside of the winter season, manatees tend to forage between six and eight hours per day, doing so either in groups or alone. Snorkelers or divers who are near feeding manatees can easily hear the sound of their chewing, which resembles the sound made by a person eating cornflakes. Foraging can take place any time of day or night, but in some circumstances (e.g., with aggregated animals at a warm-water refuge), manatees have been reported to leave the protected areas where they spend much of the day and forage actively at night.

Manatees that gather at warm-water refuges in the winter certainly interact with one another, but the driving force that causes them to come together in this situation is a response to cold weather and cold water, not social cues.

Although foraging and feeding take up much of the day for manatees, they also engage in an expected set of behaviors and activities: resting or sleeping, traveling, socializing, and mating. Whereas manatees are not generally considered to be as gregarious as some other coastal marine mammals in Florida, especially common bottlenose dolphins (*Tursiops truncatus*), manatees and dolphins have the same type of social system, referred to as fission-fusion. The term is very descriptive, because each manatee may alternate between spending time with a number of other individuals and being alone over the course of a day. Group composition may vary, but generally manatees stay in small, fluid groups, rather than by themselves. In contrast to dolphins, manatees are unaggressive.

The only long-term social association in manatees is between a mother and her calf, as they may remain together for a year or more. The mother-calf pair stay in touch with one another visually or, more often, use the high-pitched chirps and squeaks typical of the species. Manatees can communicate general information (e.g., alarm sounds) to others

in their species, but they also have individually distinctive signature vocalizations, which allow different manatees to recognize one another acoustically. Given the limitations of using vision underwater, acoustics provide the most important means of communication and threat detection for manatees.

Manatees in general, and especially females with calves, mouth and touch one another frequently. In addition, manatees exhibit play behavior, such as body surfing and follow the leader. Alone and in groups, manatees exhibit considerable curiosity and often mouth or otherwise interact with objects they find in their environment. This extends to human divers, with whom manatees may engage by clutching a diver's legs and gently rolling underwater.

When manatees travel from one location to another, they often travel single file, but calves swim next to their mothers, in part for energetic reasons, since the drag they experience moving forward is reduced when they are positioned closely adjacent to the rear part of her body. Manatees are powerful swimmers, capable of moving in bursts in excess of 20 miles (32 kilometers) per hour and of covering more than 50 miles (80 kilometers) in a day. As manatees migrate long distances to seek warm-water refuges or other resources, they tend to stop at predictable locations along the Atlantic or Gulf Coasts or the Intracoastal Waterway. The sensory mechanisms they use to navigate over long distances and select stopping points are not known, but cognition and communication expert Gordon Bauer and I have proposed that taste may be important

**In clear water, over a sandy bottom, it is especially easy to observe the behavior of manatees, alone or in social groups. Not surprisingly, the best study of manatee behavioral ecology was conducted at the aptly named Crystal River.**

The close relationship between a female manatee and
her calf is apparent in a number of ways, including a lot
of touching, vocalizing, and even playing.

(as it is for species like salmon). We think it likely that manatees use taste to detect small changes in salinity and follow a salinity gradient to freshwater to drink. We also believe that male manatees use taste to detect female manatees that are in estrus.

The reproductive behavior of manatees conforms to a strategy called scramble competition. When a female manatee becomes sexually receptive (typically other than in the winter season, which stresses the animals' energy reserves), she becomes the focal point of an estrous (mating) herd, which consists of the female and as many as three dozen attending males. The composition of a mating herd is fluid, with individual males leaving or joining the herd regularly. Although the herds can remain together for as long as seven weeks, a female is assumed to be in estrus and thus receptive to mating for only a short time—maybe a couple of days. In the meantime, the males push and shove one another, sometimes vigorously, in order to gain access to and impregnate the female. Genetic studies are being conducted at Mote Marine Laboratory (in Sarasota, Florida) and Hood College (in Frederick, Maryland) to try to identify successful sires and their attributes. One question is, are the most successful sires the largest individuals, or do smaller, quicker males have a selective advantage?

Although there are very rare reports of manatee calves being preyed upon by sharks, Florida manatees have essentially no natural predators today. In the past, subsistence communities hunted manatees, but that no longer occurs in Florida. Manatees in other parts of the world, however,

**This juvenile manatee approached photographer Wayne Lynch unsolicited and started to chew on his wrist and tug on his vest.**

A frequent and noticeable component of manatee social interactions involves individuals nuzzling and mouthing one another.

are hunted by people and are reportedly targeted by caimans and sharks; dugongs are pursued by subsistence hunters, and successful attacks on dugongs by large tiger sharks and killer whales have been reported. The daily existence of Florida manatees is pretty idyllic, however, except for interactions with an ever-growing population of humans.

When an estrous herd forms, the activity of the group can stir up considerable sediment, making the surrounding water murky. Estrous herds can remain together for more than a month and can be composed of up to three dozen males with an estrous female.

Potential predators on manatees or dugongs include large sharks (*top left*), black caimans (*top right*), and killer whales (*bottom*).

The flippers of a manatee come in handy in a number of ways. These animals can "walk" along the bottom, aided by flippers projected stiffly downward to the sediment; food can be stuffed into their mouths, using their flippers in combination with their prehensile lips and oral bristles; and their flippers can even scratch a facial itch.

# 6

# *Reproduction and Life History*

Although effective conservation requires much more than scientific information, knowledge about a species' or a population's reproduction and life history is especially useful to inform conservation decisions. Such is certainly the case for Florida manatees, although it could be argued that limitations in available funding have contributed to other aspects of manatee biology being prioritized, leaving some surprising holes in our understanding (e.g., of gestation time).

The term "life history" is loosely used, but it appears to be rarely understood outside of the fields of ecology and wildlife management. Life-history traits affect the ability of a species or population to succeed under specific environmental conditions or constraints; thus some taxa are well adapted to maintaining stable populations at or near carrying capacity over time, whereas others are suited to a rapid population expansion when resources permit, and an equally rapid demise when those resources disappear. Among the life-history traits that are most often and most usefully assessed are the following:

- age at sexual maturation
- litter size
- reproductive life span

**Accompanied by the usual consort of fish, manatees can take a nap when positioned either belly up or belly down.**

- frequency of a female's producing offspring
- age-specific survival

By most measures, manatees breed fairly slowly and produce relatively few offspring in their lifetimes. For example, female manatees generally bear a single calf every 2.5–3 years; twins occur between about 1.5 percent and 4 percent of the time. The mothers care for their young for many months and may even do so for as long as a couple of years. Females that have been severely injured, however, may not breed for several years, while their bodies try to heal.

Although the age-determination technique of counting annual growth-layer groups in the periotic bone (associated with the ear) is effective for young animals, the breakdown and reassimilation of bones (resorption) and their modification over time limit the applicability of this approach, especially for older animals. The greatest documented age for a wild manatee was 59 years, but it seems likely that manatees can exceed 60 years of age in the wild, as they do in captivity. Their age-specific survival is high, and even relatively old manatees have not been shown to experience a reproductive decline.

The most surprising exceptions to the expectations for manatees' life history and natural history are that (a) they can become sexually mature when as young as 2 years old in males and 3 years old in females (more generally, both sexes mature at around 4–5 years old); and (b) they are able to adapt to and survive well in a changing environment, one that has been extensively modified by Florida's burgeoning human popula-

Most adult manatees have an impressive girth, but pregnant females stand out physically. This pregnant manatee was photographed on 14 February, toward the end of the winter season. The peak of calving occurs outside of winter, between March and November.

Not unlike many other social mammals, such as common bottlenose dolphins, Florida manatees engage in frequent sex play. This can be especially evident in juvenile animals.

**A young manatee silently cruises along in some of Florida's clear waters.**

tion and may experience significant future restructuring as a result of the impacts of climate change (e.g., hurricanes of greater intensity).

With regard to a manatee's age at the onset of sexual maturation, even 5 years old is surprisingly young. The life span of a manatee is comparable to that of a human before advanced medical care existed, but humans generally achieve sexual maturity in their teens. A female manatee that is impregnated when 5 years old and produces a calf at age 6 and every three years thereafter (ages 9, 12, and 15), would produce four times as many offspring by age 16 as a human female of that age who becomes pregnant for the first time in her midteens. The point is that just this single life-history attribute (age at sexual maturation) may give Florida manatees a higher lifetime reproductive potential than is the case for many other large mammals, marine or otherwise. This capacity positively influences their potential for recovery from population declines.

Modeler and statistician Mike Runge and his collaborators have developed a population model based on the life history of Florida manatees. They have suggested that the relatively small manatee subpopulation occupying the St. Johns River has grown at a rate of 6.2 percent per year.

A rather large calf acquires some extra nutrition by nursing. Even after a calf is nutritionally independent and is feeding well on vegetation, it is not uncommon to see an occasional bout of nursing. The mammary glands of manatees are located in the armpit region, rather than along the abdomen, as is the case with many other mammals.

Calves are nutritionally dependent on their mothers' milk immediately after birth, but within a matter of a few months, the babies begin to supplement this milk by adding plants to their diet. Approximately two-thirds of the calves stay with their mothers during just one winter season. Rarely, calves may remain with their mothers and be nursed for as long as twenty-four months.

This figure is higher than that of other manatee subpopulations (often referred to as management units) in Florida, but it corresponds closely with the theoretical maximum population growth rate of 6.6 percent per year for Florida manatees calculated by Helene Marsh, when she assumed a 2.5-year interval between calves and a mean female age of 4 years when first giving birth. These assumptions may not be conservative, but they are plausible and illustrate that in a rapidly changing Florida environment, manatee life-history attributes give this subspecies a fighting chance for success. The figures also suggest that for at least some subpopulations of manatees, carrying capacity has not yet been reached.

Manatees frequently bask at the surface. At such times, an unwary manatee may be especially vulnerable to collisions with fast-moving boats.

# 7

# *Habitat Protection*

My first and perhaps most enduring demonstration of the toughness of Florida manatees came while I was in graduate school in the mid- to late 1970s at the University of Miami's Rosenstiel School of Marine and Atmospheric Science. I worked with and for Dan Odell, who, among other things, catalyzed the creation of a marine mammal stranding network for the southeastern United States. I supported myself in part by assisting Dan to recover and perform necropsies (animal autopsies) on the bodies of dead manatees and cetaceans.

One day we brought in a manatee carcass that made a real impression. A boat had struck the animal while it was alive, breaking some of the ribs due to impact and slicing deeply between them with the propeller. The manatee died, I recall, with part of one lung sticking outside its body, but with a mouthful of well-chewed food and more food in its stomach, suggesting that it actually lived—and foraged—for a time with its horrid injuries.

On a more regular basis, these animals' toughness and will to survive are demonstrated whenever I see severely scarred and deformed manatees surviving year after year, and even producing calves. I can only imagine the pain and trauma they suffer from such injuries. In Sarasota waters, two venerable females named Manx and Victoria were

It is not uncommon to observe mutilated or even severed flukes. When manatees detect an oncoming boat by the sound of its engine, they often attempt to dive. In such a scenario, the fluke is the last part of the manatee to leave the surface, resulting in severe and more-frequent injuries to that body part.

studied for decades by my staff at Mote Marine Laboratory and were noteworthy not only for their multiple and distinctive scar patterns but also for their persistence, until they were ultimately killed in collisions with boats.

I have since learned from my veterinary friends (including Greg Bossart, Ray Ball, Andy Stamper, and Mike Walsh) that immunologically, manatees are like tanks, able to fend off attacks by disease-causing agents that could destroy a mere human. Wild animals from a wide range of species tend to be pretty tough (compared with humans), but manatees can sustain a lot of damage and still remain alive.

In addition to being able to survive physical assaults and insults better than many other animals, manatees consume a wide variety of plants in a broad range of salinities (~0–35 parts per thousand) and can thrive both in close proximity to people (e.g., downtown Miami and Tampa) and in Florida's wild places (e.g., the Everglades). I have likened manatees to coyotes, because both species show an uncanny ability to survive—and even thrive—around people.

As a result, Florida manatees occupy the awkward position of being an endangered subspecies that is probably at or near the carrying capacity of their Florida environment, which includes certain human activities (e.g., boating) and unpredictable random events (e.g., red tides and cold fronts) that can lead to a considerable number of illnesses and even deaths for this subspecies. Hunting pressure and the incidental killing of manatees by their becoming tangled in fishing gear are minimal for manatees in Florida; rather, Florida manatees, resilient and adaptable

as they are, may ultimately be sustained or lost due to issues associated with their habitat.

Which habitat features, then, are essential to Florida manatees? Some stakeholder groups ask additional questions, such as how many manatees can and should the Florida coastal ecosystem support (i.e., what is its carrying capacity, and when do we have "enough" manatees?), and have we reached the point where we have so many manatees that the environment is suffering? Perspectives on these and related issues are the focal points of this chapter.

West Indian manatees have been shown to consume an impressive 100 or so genera of plants in a range of marine, brackish, and freshwater habitats, so unless plants are physically removed because of development projects or dredging, silted in by sediments suspended in the water by the passage of boats or by hurricanes and storms, or exposed to water with a high level of contaminants or other degrading factors, one would expect that manatees would have plenty to eat. In many parts of Florida, this assumption would hold up well, but the 43 percent reduction of

Even some very young manatees bear scars resulting from a collision with a boat. An assessment of the scar patterns of adult manatees suggests that some individuals may be struck several dozen times during their lifetimes. At present, more than 3,800 Florida manatees, individually identifiable due to their unique scar patterns, exist in a statewide computerized database called MIPS (Manatee Individual Photo-Identification System).

In the Crystal River area, there are a number of separate freshwater springs. One of the springs used most by manatees in winter is the beautiful Three Sisters Spring.

seagrass coverage over just four years in Brevard County, an area used by 2,000 or so manatees, underscores the vulnerability of this critical habitat resource. The seagrass decline there was attributed, at least in part, to algal blooms, more widespread eutrophication (when excessive nutrients, particularly phosphorus, result in the depletion of dissolved oxygen in the water), generally diminished water quality, and reduced light penetration (thus slowing plant growth). As 2016 got underway, a new algal bloom event appeared to be taking place, which could further diminish the already-depleted seagrass beds of Brevard County.

Aside from food availability, there are several other habitat needs for manatees. The issue of appropriate habitat for manatees was a key question when Dana Wetzel and I offered our thoughts on the suitability of the Grand Cul-de-Sac Marin for the Guadeloupe manatee reintroduction project. The features we considered included the availability of adequate food (especially seagrasses), access to freshwater to drink, protected areas (especially for calving and rearing their young) that were away from the effects of boating and hurricanes, and the presence of clean water. Since

Guadeloupe and nearby areas in the Caribbean Sea do not experience cold winters, warm water (a major habitat requirement for manatees in Florida during the winter) is plentiful and therefore not an issue of concern. We still have much to learn about the habitat requirements for manatees, but if the will to do so exists, these resources can be managed for the benefit of both people and wildlife.

Perhaps the greatest enemy to the success of habitat conservation in Florida—and globally—is a burgeoning human population. Some studies project that the number of people living in Florida will increase from around 20 million in 2015 to 26 million in 2030, an astounding 30 percent increase in just fifteen years. Globally, the human population is anticipated to grow from 7 billion in 2011 to 10 billion in 2083, a monstrous 43 percent increase in only seventy-two years. Even with creative and proactive planning and mitigation, how will our planet—and the living resources and wild ecosystems on it—persist in the face of the expanding numbers and requirements of one very demanding and egocentric species, *Homo sapiens*?

### Warm Water and Manatees

Much has been made of the relationship between manatees and warm water in Florida. Survey efforts take advantage of the fact that manatees gather in large numbers at known warm-water refuges around the state. The most recent Florida Manatee Recovery Plan was published in 2001 by the US Fish and Wildlife Service. At that point, the document noted that there were eleven primary and twelve secondary warm-water ref-

At Crystal River, some areas are forbidden to divers and swimmers, in an attempt to give the manatees space and a time when they can be unmolested, if they wish. Even in the unregulated areas, it is possible to see manatees "just being manatees" in the clear waters of various springs.

Hunter Spring, Crystal River, provides a wonderful habitat for manatees and other species. The explosive growth of Florida's human population, however, will make it increasingly difficult to conserve both the environment and the species that need that environment.

uges in Florida, but its list of sites and their designations as primary or secondary areas is seriously out of date. A complete list of vital warm-water resources for manatees would probably include more than three dozen locations today.

Surprisingly, despite their acknowledgment that manatees in Florida require access to warm water in winter, scientists and managers do not have a good understanding of how manatees are able to control their body temperature (i.e., thermoregulate), and questions have arisen about whether manatees occupy warm-water refuges because they require the warmth, or just because they enjoy it. Daniel S. (Woody) Hartman, a noted pioneer in the field of manatee biology and conservation (and a Darwinesque-quality observer of animal behavioral ecology), went so far as to state in the 1970s that "except during unseasonably cold weather, manatees bask in tepid springs less out of thermoregulatory necessity than for the salubrious sensation. The value of springs and other warm-water sources as refuges from cold is probably restricted to protracted periods of freezing or near-freezing temperatures, which are uncommon

1-Crystal River/Kings Bay
2-Homosassa Springs
3-Apollo Beach Power Plant
4-Matlacha Isles
5-Ten Mile Canal
6-Fort Myers Power Plant
7-Port of the Islands
8-Fort Lauderdale Power Plant
9-Port Everglades Power Plant
10-Riviera Power Plant
11-Canaveral Power Plant
12-Blue Spring State Park

in Florida." Hartman was an uncanny observer of manatee behavior and ecology, and many of his conclusions still stand today, forty to fifty years after he conducted his studies. Nonetheless, he clearly underestimated the importance of warm-water refuges for manatees in winter.

Hypothermia (a lowering of body temperature) results when the systems the body uses for heat production and conservation are overridden by exposure to cold. Ultimately, hypothermia can lead to death. The bodies of humans and many other mammals possess several mechanisms that they use to stay warm during cold weather. The presence of fatty (adipose) tissue and hair provides some insulation. Circulatory changes allow heat to be maintained within core areas of the body, in part by preventing warm blood from reaching peripheral areas, where heat would easily diffuse to the environment. Shivering consists of a series of rapid muscle tremors, which use metabolic (chemical) processes to produce heat; shivering can actually increase an individual's metabolic rate to a level about five times its resting value. Exercise makes use of larger muscle movements and creates even more heat in an individual. Tests with monkeys show that long-term exposure to nonlethal levels of cold can lead to a 30 percent increase in the metabolic rate of an individual, presumably through increased secretion of a particular hormone (thyroxin) by the thyroid gland. Other responses to cold temperatures by humans and other mammals can include migrating and basking.

The mechanisms manatees use are presumed to be generally similar to those employed by humans and other mammals. Rudy Ortiz, Graham Worthy, and Duncan MacKenzie have shown that the metabolic rates

The map shows the locations of some of the more important warm-water refuges for manatees. The formal names of some locations have been abbreviated to allow them to fit on the image.

On a cold winter morning, steam can come off the surface of the water at the Chassahowitzka River warm-water refuge. An anhinga perches on a limb, drying its wings after diving for fish to eat.

A manatee, attended by redbreast sunfish, rests in a warm-water refuge.

of cold-stressed manatees rise, although the physiological mechanism for this increase is not understood. Cold-stressed manatees may migrate to warm waters (and migration involves exercise), and they bask at the surface on sunny days. Although their overall metabolic rate is lower than expected, they may have a couple of mechanisms that help compensate for this deficiency: (a) they have effective vascular counter-current heat exchangers in their armpit areas (axillae) and the base of their fluke (caudal peduncle); and (b) the fermentation of plant material in their enormously enlarged hindguts probably provides an important metabolic source of heat.

When an individual is exposed to extreme cold, the circulatory modifications that promote the maintenance of warm core temperatures expose peripheral regions of the body to degrees of cold that can lead to frostbite and tissue death; this happens in both humans and manatees. In manatees, death from exposure to cold may be manifested in two general ways.

Long-term exposure to moderate cold may lead them to stop eating, which may have several effects, including reducing their energy stores and their body's insulation, as well as lessening the amount of heat provided by fermentation in their gut. Manatees suffering from such chronic exposure exhibit a lack of fat reserves, a general bodily wasting away (known as cachexia), and empty gastrointestinal tracts, along with skin lesions. Animals exposed to excessive or prolonged cold may also possess greatly enlarged lymph nodes near their armpits. In captive manatees, feeding becomes irregular at temperatures below 64°F–66°F (~18°C–

On cold days when the sun is out, manatees often bask at the surface to benefit from the heat that the sun provides. Not surprisingly, such days are excellent times to conduct aerial manatee surveys, since the animals are visible and relatively easy to count, even in murky water.

Florida manatees, such as this visitor to a warm-water refuge, have rather rough skin, which facilitates the attachment and growth of organisms. Dugongs and Amazonian manatees have much smoother skin, similar to that of dolphins.

19°C), a range that corresponds well with the point at which thermal stress has been suggested to occur, and feeding and other activities cease at water temperatures below 61°F (~16°C).

On the other hand, manatees may die quickly from exposure to intense cold, in which case a diagnosis as to the cause of death may be extremely difficult. In humans, uncoordinated contractions of the heart muscle (myocardial fibrillation) can cause death in individuals whose body temperature drops below 82°F (~28°C). Also, diseases such as pneumonia or bronchitis may occur in cold-stressed manatees, as they do in humans.

A confounding variable when dealing with the effects of cold on manatees involves individual variations among these animals. Just as some people are better able to withstand cold weather due to their body's insulation, proportions (the surface-area-to-volume ratio), resting metabolic rate, health status (including that of their circulatory system), nutritional status, or other variables, the same appears to hold

true for manatees. Some manatees seek warm water in October, at the first sign of cold, whereas others may rarely, if ever, visit a warm-water refuge.

Even given a lack of empirical data and the variability among individual Florida manatees, some clear conclusions can be reached about their responses to cold. A number of pathological changes are associated with cold stress syndrome in manatees, symptoms that include emaciation, the depletion of lymph node tissues and fat stores, a reduction in the size and number of fat cells (serous atrophy), heart deterioration (myocardial degeneration), inflammation of the intestines (enterocolitis), an increase in the number of skin cells (epidermal hyperplasia), and pus-filled areas on the skin (pustular dermatitis). Manatees affected by cold stress syndrome may die, due to a lack of nutrition and the suppression of their metabolism and immune functions, as well as from secondary diseases. Inflammation and infection routinely lead to skin lesions (diseased places), which can occur anywhere on the body but are most prevalent on the head, flippers, and fluke. Manatee observers typically "score" (assign a rating to) such lesions as a reflection of the severity of cold exposure and cold stress in individual manatees. A range of findings that occur at necropsies on manatees suggest a cascade of reactions in which the effects of malnutrition and cold have a negative impact on the animal's immune system, which in turn leads to disease-caused changes in its skin and other organs, as well as opportunistic infections (ones that normally are harmless but cause problems when

Manatees that are exposed to cold can develop a set of conditions called cold stress syndrome. Skin lesions, or diseased areas which manifest themselves as thickenings on the outer layer of skin, are one of the features associated with this syndrome.

Some manatees show great site fidelity, returning to the same warm-water refuge every winter. Managers must be sure to plan ahead and ensure that a network of reliable warm-water sites is available in perpetuity for Florida's manatees.

the body's resistance is impaired). There is also evidence that normal immune functions sometimes are not regained quickly in cold-stressed manatees, even after extensive rehabilitation of these individuals, suggesting that the consequences of severe cold stress may affect a manatee's health and survival for months or even years.

In 2010, due to the extremely cold winter of 2009–2010, 282 manatees were reported to have died in Florida as a result of cold stress; this represents 37 percent of the total mortality for manatees in Florida that year. Manatees of all size classes die from cold stress syndrome, but subadult manatees tend to be disproportionately represented. It is very important to understand that the lethal impacts of exposure to cold can remove significant numbers of manatees from the population in a given year. Long-term sublethal effects from extreme cold on the animals' immune functions or reproductive potential, however, have not been thoroughly examined, but these may be significant with regard to the long-term population status of manatees.

## Case Studies: When the Warm Water Ceases

What happens to manatees when the warm-water flow ceases in locations where the animals have been conditioned to expect it? Studies of tagged manatees show considerable year-to-year fidelity in terms of the natural and artificial warm-water sites they use along the Atlantic coast of Florida, and a number of photo-identification studies around the state have confirmed this behavior. There have also been some "experiments" that document the extent of the tenacity (or, perhaps, the lack of adaptive behavior) on the part of at least some manatees.

*Ft. Myers Power Plant.* In January 1985, the warm-water discharge at the Florida Power & Light Company's plant in Ft. Myers was stopped for both economic and maintenance reasons. Despite the termination of this warm-water flow, over 300 manatees remained in the area of the plant during very cold weather. Many moved a short distance to the Franklin Locks, up the Caloosahatchee River, until the heated discharge was restored, at which point they returned to the region of the plant itself. The manatees did not appear to follow a thermal gradient from cold water to warmer water; rather, they moved through colder water to reach the warm water at the plant, which could have created considerable stress if the warm-water discharge had not been resumed. Luckily, no cold-related manatee deaths occurred during this event. To prevent a similar situation from recurring, the Florida Power & Light Company installed artesian wells at this plant prior to the following winter, allowing warm water to be pumped into the discharge canal, even when the plant was not operating. These wells were abandoned by the company after its Ft. Myers plant was modernized and more-suitable options for providing warm water became available from the repowered units.

*Jefferson-Smurfitt Plant.* In winter 1997–1998, the discharge at the Jefferson-Smurfitt Corporation's Fernandina Beach power plant was diffused, thereby eliminating a long-term warm-water source in extreme northeastern Florida. Despite the onset of cold weather, not all manatees moved south. Some animals remained in the area of the plant, although they made local exploratory trips, apparently seeking alternative warm-water sources. The number of manatees affected by the event was small, but between October 1997 and March 1998, a large percentage of the animals in that region died or were taken into rehabilitation facilities for treatment. The cause of death was not determined in most cases, but cold stress was thought to have been a problem in at least some of the deaths and rescues, even though that particular winter was relatively mild.

At present, manatees that are safely within the boundaries of one of Florida's protected warm-water refuges are relatively shielded from both natural and human-induced sources of stress. Existing stressors for manatees (including but not limited to motorboat collisions, noise and chemical pollution, and red tides), however, are likely to be multiplied in the future, due to dramatic growth of the human population, a possible loss of some warm-water refuges, and the effects of climate change.

*H. D. King Power Plant.* This power plant, located in Ft. Pierce and operated by the Ft. Pierce Utilities Authority, was essentially shut down in winter 1997–1998, when it operated for only sixteen days. This plant was eventually decommissioned in 2008 and demolished in October 2009. Nonetheless, the former discharge area continued to attract manatees every winter for several years. On 27 January 2000, during cold winter weather, twenty-six manatees were observed at the H. D. King Plant, despite the fact that, at least when the aerial survey occurred, there was no warm water being discharged.

*Riviera Plant.* In February 2000, a combination of routine maintenance and a minor fire created by an oil leak caused this Florida Power & Light Company plant to be shut down at a time when large numbers of manatees were present. The ambient water temperatures remained above 61°F (16°C), so the company did not have to initiate provisions in their Manatee Protection Plan. The plant did not run throughout February, but air and water temperatures rose within a week of the shutdown. Aerial counts showed that many of the manatees present prior to the shutdown simply stayed in the general area of the plant (in Lake Worth Lagoon or in nearby Hobe Sound) for several days, although some may have moved south to plants in Ft. Lauderdale. No manatee deaths that could be confirmed as cold related occurred during this event.

*Lauderdale Plant.* The discharge zone for this Florida Power & Light Company plant is extremely sheltered from human disturbance and, in theory, would appear to be a good place for manatees to seek a warm-

This manatee, occupying a warm-water refuge, exhibits what pioneering manatee expert Woody Hartman called a "stretch posture."

water refuge. Although the plant had once provided shelter for up to fifty-two manatees in winter, its inefficient and sporadic operation, and the subsequent low output of warm water, relegated it to a relatively insignificant status as a warm-water refuge by the late 1980s. The plant was closed in October 1992 for modernization, to increase both its efficiency and production capacity. After the modernization was completed, the plant operated with increasing frequency and was sought as a warm-water refuge by greater numbers of manatees. By 1999–2000, more manatees were observed in the discharge zone of this plant than at the nearby Port Everglades plant for the first time (124 at Lauderdale vs. 111 at Port Everglades on 27 January), a trend that has continued to the present, with an all-time high count of 949 manatees at the inland Lauderdale plant on 17 January 2012. What has been documented at this plant shows that the importance of certain locations for manatees can change dramatically over time and provides some empirical data on the timing of transitions.

For more than forty years, manatees have aggregated in Kings Bay, Crystal River. In winter, single-day counts of manatees in Crystal River have numbered close to 800 animals—a lot, to be sure, but less than a couple of single-day counts at two of the Florida power plants.

## The Uncertain Future of Warm Water for Florida Manatees

It is increasingly apparent that manatees in Florida require a network of resources to survive. That network must include daily or frequent necessities, such as food, freshwater, and clean water, but it must also involve warm water in winter. Unfortunately, many of the primary warm-water refuges are likely to have finite life spans as costs of fossil fuels escalate, due to diminishing reserves, and the availability of an adequate flow from springs is affected by an aquifer that is being overtapped for agricultural and other human uses.

For nearly two decades, state and federal agencies (the Florida Fish and Wildlife Conservation Commission and the US Fish and Wildlife Service) with oversight for manatees and their conservation in Florida led a somewhat humorously named Warm Water Task Force, charged with developing mitigation options and facilitating their implementation, in order to keep adequate warm-water resources for manatees available forever. That group, however, has not met in many years—and the issue remains unresolved. Similarly, but more broadly, under the US Endangered Species Act, the US Fish and Wildlife Service is statutorily charged with developing a Manatee Recovery Team and Plan to formally ensure that necessary recovery actions (including the provision of warm water in winter) are implemented; the most recent Manatee Recovery Plan is dated 2001, which is also the last year that the Recovery Team met.

Starting in 2010, the Florida Power & Light Company initiated the sequential modernization of three power plants along Florida's east coast. In the process, old, inefficient oil-burning plants were converted to more-efficient and cost-effective natural gas–burning plants. Permission to make the conversions hinged on a commitment by the power company to sponsor meetings and otherwise promote the development of mitigation options to ensure that manatees dependent on industrial warm-water sources would have reliable warm-water alternatives in perpetuity. That commitment by Florida Power & Light Company is, however, dependent on the state and federal agencies involved with the Warm Water Task Force completing an agency-approved plan of action. Fortunately (for both the company and the manatees), Florida Power & Light had the vision to contract proactively with me for a report that outlined several options, plus the mechanisms to achieve those options, in order to create a network of nonindustrial warm-water refuges for Florida manatees. This report provided an overview of locations along Florida's east coast that could be developed and protected as warm water refuges for manatees if existing power plants were retired. With significant help from spatial analysts with the Florida Fish and Wildlife Conservation Commission, I based my recommendations for future warm-water sites on several parameters:

- proximity to seagrasses
- easy access to freshwater
- good water quality
- low levels of noise and disturbance by boats
- low levels of human-related manatee deaths
- proximity to migratory corridors
- proximity to summer habitat
- locations south of the Sebastian River (approximately the historic northern winter boundary for manatees)
- relatively low human population size and projected growth
- availability of land for the erection of warm-water generating equipment (e.g., solar panels)
- high levels of existing habitat and species protection
- substantial access of adjacent land areas for education and outreach facilities and programs

Florida may lack the dramatic beauty of some other parts of the earth, but it is certainly romantic to see an image of moss-covered trees; calm, clear waters; and a gentle, gigantic manatee gliding along in a warm-water refuge.

People have sometimes commented on the extent to which manatee adaptations and behaviors may be paradoxical. Their use of warm-water discharges seems to be another seemingly contradictory situation. Over time, manatees show a great ability to learn and utilize resources, such as artificial warm-water refuges, that did not exist historically. On the other hand, they may react to the sudden elimination of such resources by simply waiting for the resources in question to reappear. Given that tendency, in the event that traditional or current warm-water sources

disappear, proactive contingency plans must provide manatees with easy access to a network of alternative resources. Stated otherwise, if a system of new, nonindustrial refuges is created, it must be operational well in advance of the loss of existing sites. In addition, the new refuges must be placed so that manatees presently relying on existing refuges can find and *very quickly* learn to use the new locations. Today, proactive planning to perpetually ensure adequate warm-water resources for manatees in winter must be a top priority for conservationists and management. Sadly, informed and achievable solutions remain clouded, at least into the foreseeable future.

### Freshwater Consumption and Need

Noted manatee biologist Tom O' Shea stated many years ago that manatees and freshwater are like children and ice cream: even though children do not need ice cream to survive, they certainly enjoy consuming it. Tom suggested the same scenario for manatees drinking freshwater. Biologists, including O'Shea, have used the manatees' inclination to drink freshwater from a garden hose as a not-too-sophisticated lure to attract free-ranging manatees to a particular spot where they could be captured, measured, and have samples drawn for health assessments. Like some other aspects of manatee biology, the animals' specific capabilities and limits with regard to regulating their internal water and salt balance (osmoregulation) and their need for freshwater remain a little hazy, but it currently does appear that West Indian manatees *do* require periodic access to freshwater.

When manatees occupy marine waters for any length of time, they can acquire a growth of barnacles, primarily on their backs and flukes. The barnacles fall off soon after a manatee returns to a freshwater habitat.

Like Florida manatees, American alligators occupy fresh and brackish waters, and these two species may cross paths regularly. Although some alligators reach enormous size, there is no evidence that they prey on manatees in Florida.

The scant literature on manatee osmoregulation indicates that the shape and structure of a manatee's kidneys seem to be adapted to produce very concentrated urine (similar to the capabilities of desert rodents), suggesting that manatees can exist for some time without access to freshwater to drink. The presence of manatees with large barnacles on the surface of their skin reinforces the length of time that some manatees remain in marine environments.

Physiological studies of captive manatees, however, suggest that at least periodic access to freshwater (e.g., every few days) is important for the animals' health and body condition. For example, captive manatees held for nine days in seawater and fed marine plants could maintain a specific level of salts in the fluid around their cells (electrolyte homeostasis), but they lost weight, increased their salt concentrations (plasma osmolarity), and ate less. Based on these physiological findings, coupled with the movements and habitat use of tagged manatees, the prevailing scientific opinion is that manatees deprived of freshwater for more than a few days (maybe just four to five days) can tolerate this situation by breaking down (metabolizing) their body fat and producing water as a byproduct of this process, but they physiologically need freshwater to survive in the longer term. Thus maintaining access to predictable sources of freshwater for free-ranging Florida manatees is an important conservation consideration.

## Water Quality

Manatees, or any other organisms, thrive best in habitats that have minimal pollution, debris, and other sources of contamination. Florida manatees are extremely well studied, but most of the information available through science is focused in a few disciplines (ecology, behavior, population dynamics), leaving some relative black holes in other areas.

When considering water quality, there are several possible contaminants, or stressors, that should be assessed in relation to both their lethal and sublethal effects on manatees. In addition, changes in water quality can influence events, such as red tides, that can have indirect but devastating short- and long-term impacts on local manatee populations.

The chemical contaminants that seem most relevant to manatee health and well-being are the especially dangerous organic ones: organochlorine pesticides (containing chlorinated hydrocarbons), famously including DDT and DDE, kepone, and other products that have allowed Florida's agricultural empires to remain vibrant; the most toxic components of oil and gas pollution, called polycyclic aromatic hydrocarbons (PAHs); the persistent and highly toxic polychlorinated biphenyls (PCBs); and the deadly new class of compounds called polybrominated diphenyl ethers (PBDEs) which are present as flame retardants in our computers, furniture, and clothing. All of these classes of compounds are persistent in the environment and in animal tissues. None have been well studied in manatees, although research regarding other mammals (including controlled experiments using laboratory animals) shows both lethal effects and long-term sublethal effects on reproduction and longevity.

Recent catastrophes associated with oil spills suggest that sirenians and other marine mammals may be especially vulnerable to such events. For example, the Nowruz oil field spill in the Arabian Gulf (1983–1984) killed more than 150 dugongs and may have impaired the health of far more individuals. More recently, exposure to hefty doses of oil during the Deepwater Horizon oil spill in the northern Gulf of Mexico (2010) coincided with immediate deaths for a number of common bottlenose dolphins, as well as impaired immune functions, body conditions, reproduction, and longevity for others.

Marine mammals tend to have large fat depots in their bodies, which predisposes them to accumulating organic chemicals over time (a pro-

Manatees are said to have a fusiform (spindle-shaped) body, tapered at either end. Though not as streamlined as some other species of marine mammals, a manatee's fusiform body, powered by a broad fluke, allows it to move through the water far faster than the fastest human swimmer.

cess called bioaccumulation). Male marine mammals are especially susceptible to bioaccumulating organic contaminants and suffering consequences from it, because they lack a mechanism for effectively ridding their bodies of such chemicals. On the other hand, mature females lose some of these bodily burdens through their milk, which is rich in fats (lipids). In some well-studied species, such as common bottlenose dolphins, the offloading of maternal contaminants into both milk and offspring is claimed to be why older females have lower levels of organic contaminants in their bodies than older males, as well as why first-born dolphins rarely survive. This may actually be an issue for Florida manatees, since roughly 20 percent of all mortality for them is perinatal, referring to near-birth infants for which no cause of death is currently ascribed.

Nutrition is one factor in favor of manatees with regard to the bioaccumulation of organic contaminants in their tissues. Herbivores tend to ingest and store fewer contaminants than do similar-sized carnivores. Species at the very top of the nutritional (trophic) pyramid (e.g., polar bears and killer whales), which eat other carnivores, tend to be the species at greatest risk for bioaccumulating organic contaminants.

In Florida, another class of contaminant is gaining attention: pharmaceuticals. The increasing presence of drug-related chemicals in coastal waterways is cause for some alarm, especially in highly developed coastal regions of Florida (Miami-Ft. Lauderdale and Tampa-St. Petersburg). One of the more insidious effects of such chemicals is the potential

disruption or impairment of reproduction in nonhuman mammals, like manatees and dolphins, through increasing levels of compounds that mimic estrogen (xenoestrogens), such as those in birth control drugs. The list of known chemicals that can cause males to take on feminine characteristics includes nonpharmaceutical compounds such as DDT, which has been implicated in effects on male alligators (*Alligator mississippiensis*) in Florida. Collectively, xenoestrogens have been shown to impact male development in every class of vertebrate animals (mammals, fish, birds, reptiles, and amphibians), yet the majority of the over 100,000 recently introduced chemicals are underregulated.

Perhaps the best-substantiated link between pollution and its effects on manatees involves runoff from fertilizers, which can promote rapid eutrophication and algal blooms. Notorious among such events are red tides, which are caused by the explosive growth of *Karenia brevis*, a toxic dinoflagellate (a type of marine plankton). Red tides have probably occurred throughout human history, but runoff in agricultural areas has been suggested to have enhanced the frequency, intensity, duration, and toxicity of these events. Red tides in Florida wreak havoc on local fish, seabird, dolphin, and manatee populations; these events can even hospitalize and otherwise debilitate people in coastal communities.

Manatees take in the red tide toxins either by breathing the aerosol produced when the dinoflagellate cells break apart (lyse) during storms, or by eating filter-feeding organisms, such as sea squirts (tunicates), that are attached to seagrasses and can concentrate the toxins in their tis-

In Florida, common bottlenose dolphins are often found in proximity to manatees. The seagrasses and other vegetation that manatees consume are nursery areas for the small fish and invertebrates that dolphins eat. Manatees and these dolphins tend to ignore one another.

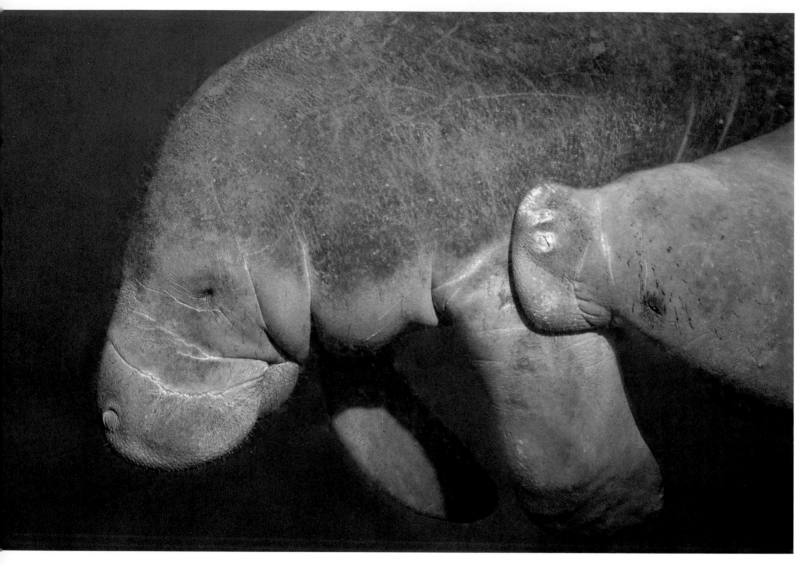

When baby manatees nurse and consume their mothers' milk, they can also ingest organic contaminants, including, but not limited to, pesticides and flame retardants. These chemicals enter the mother through the food she eats, and they are selectively found in her fat-rich tissues, such as those that produce milk.

sues. Whereas the effects of oil and other pollutants on manatees remain somewhat uncertain, red tides have been documented to kill at least 150 manatees in Florida in a single year. As is the case for exposure to cold, exposure to red tide toxins (collectively called brevetoxins) can have long-term effects on immune functions in manatees. Thus manatees that have been exposed to a major red tide event and survived may later succumb to a lesser red tide or another relatively minor stressor in the future. Some manatees reported as dying due to a collision with a boat actually displayed a biomarker indicative of exposure to brevetoxins. In such cases, this exposure may have sufficiently debilitated the manatee to predispose it to being struck by a high-speed boat.

Finally, eutrophication due to nutrient buildup can cause nontoxic algal blooms that reduce the penetration of light in coastal and inshore waters. Such is the case in the die-off of seagrasses in Brevard County. In addition, the loss of seagrasses, which can stabilize sediments, can lead to enhanced turbidity and reduced light penetration. When light

cannot reach the submerged seagrasses, photosynthesis cannot occur, and the essential forage plants for manatees will disappear.

In summary, waterborne pollution has been documented or suspected to have significant effects on manatee survival, immune functions, health, and reproduction. Keeping Florida's waterways as clean as possible is a necessary ingredient in maintaining vital habitats to sustain manatees—as well as people and other living resources—into the future.

## *Quiet Waters*

Anecdotal evidence indicates that manatees seek quiet areas in which to give birth and nurture their newborn calves. Quiet areas from which boats are prohibited are used disproportionately by female manatees and their calves in Sarasota Bay. Scientists have noted that manatees selectively use habitats in which the surrounding (ambient) noise from boats and other sources is minimal. Specifically, in seagrass beds of equal

Red tides represent a danger for manatees and many other species (including humans) in Florida. The toxins associated with red tides certainly cause the death of fish-eating bird species, such as the osprey (*top left*), bald eagle (*top right*), double-crested cormorant (*bottom left*), and common loon (*bottom right*).

North of Crystal River, some relatively undeveloped coastal habitat exists for manatees and other species. Even here, though, the stacks of a nuclear power plant serve as a reminder that humans are literally not out of the picture.

density and species composition, manatees preferred those beds with lower ambient noise (at frequencies below 1 kilohertz).

## Pending Concerns

Manatees are adaptable and tough, but many experts feel that loss of habitat may be the single factor that impacts them most into the future. To maintain Florida manatee populations, and indeed to support many human activities and stakeholder values, wildlife managers and conservationists must develop proactive plans to encourage healthy seagrass beds, ensure minimal pollution and otherwise promote water quality, protect and ensure sustainable sources of freshwater and warm water, and retain or create quiet and protected areas. To fail to do these things represents possibly the greatest threat to Florida manatees for the future.

Manatees can live for at least fifty-nine years in the wild. During that time, particular individuals may interact with each other hundreds or thousands of times. They may be the equivalent of old friends in our own species.

Manatees are powerful swimmers, sometimes covering
ten, twenty, or thirty or more miles per day. When they
are active, they breathe far more frequently than when
they rest—just like people.

# 8

# *Conservation Threats*

There are at least 6,500 manatees in Florida today. Whereas some stake-holder groups have claimed that that is too many, it could also be asked, is that enough to sustain the subspecies in the face of a growing human population and myriad natural and human-related threats?

Although historical manatee population sizes are completely un-known, and current and recent ones are not precisely or necessarily accurately known, most scientists and managers acknowledge that there are more manatees in Florida today than at any point in the recent past. Credit is due to the regulatory agencies (notably the US Fish and Wildlife Service and the Florida Fish and Wildlife Conservation Commission), the oversight agency (US Marine Mammal Commission), concerned nongovernmental organizations (such as the Save the Manatee Club), and, most of all, the people of Florida.

For management purposes, Florida manatees are considered to in-habit four management units, denoted by location: Atlantic Region, Upper St. Johns River Region, Northwest Region, and Southwest Region. Manatees within each of these units (also referred to as subpopulations) mix genetically with manatees from other units, so the designations do *not* indicate that each unit is reproductively isolated. In part, the desig-nation of regional manatee subpopulations reflects the understanding

**Manatees cannot exactly wrestle, but they certainly
enjoy rolling around with one another.**

that the status of and threats to manatees differ geographically to some
extent.

The largest management units are the Atlantic and Southwest regional groups. Based on aerial counts and adult survival estimates derived
from long-term photo-identification studies, the former population is
considered to have been increasing at a modest rate over the past two
decades. The latter unit has been thought to have been stable or in slight
decline (as a result of consistently high levels of watercraft-related and
red tide deaths), but recent reanalyses suggest that the Southwest Florida
subpopulation is stable or slightly increasing. The two smaller management units are clearly growing at a rapid rate (as high as 6.2% per year)
that approaches the theoretical maximum for the species.

The concern for me is whether we have identified and controlled
future threats that could negate the recent population growth of the
Florida manatee population. I am not convinced that we have, and, as
a result, I continue to (a) worry about the future of Florida manatees,
despite the recent positive signs for population growth, and (b) encourage a precautionary approach to management and the proper placement
of the burden of proof.

### Direct Threats to Florida Manatees

A number of publications discuss documented or putative threats to
Florida manatees. Some of the more serious issues for manatees involve
habitat attributes, such as the availability of warm water in winter; freshwater to drink; unpolluted water; adequate forage (which is affected, in

part, by water quality); quiet locations, especially for calving and rearing their young; and documented or likely effects of red tides, cold weather, and contaminants.

Some of the other significant threats to Florida manatees include collisions with watercraft, underwater noise and other disturbances (mostly associated with boats and divers), and various effects of climate change. Although certain other factors—such as entanglements in fishing gear and infectious diseases—are of some concern, the former generally results in more rescues than deaths, and the latter appears to a relatively infrequent event, at least at present.

Perhaps the best-documented cause of manatee mortality since records were kept (starting in the mid-1970s) has been collisions with watercraft. Approximately half of such fatalities are due to the impact of the collision, which can cause massive trauma, including broken bones; the other half are caused by cuts from the propeller. Such cuts are not insignificant. In some instances of collisions with very large vessels, such as barges, adult 1-ton manatees have been cut completely in half.

Blue Spring State Park, along the St. Johns River, provides an important warm-water refuge for manatees in northeastern Florida. Park Ranger Wayne Hartley is now retired, but he spent many years documenting which particular animals came back to the spring every winter.

The Gulf Coast of Florida, such as this site opposite Homosassa Springs, provides some quiet areas where manatees forage.

The accidental striking, wounding, and killing of manatees by watercraft was first noted in the 1940s. Since that time, boats have become faster and more powerful, and their designs allow many of them to enter shallow waters frequented by manatees. In addition, as a corollary of the unprecedented increase in Florida's human population, the number of registered boats in the state exceeds 1,000,000, and this figure does not include out-of-state vessels that enter Florida waterways, especially during the winter tourist season. This combination of features, coupled recently with an increasing manatee population in the state, means that the frequency of collisions, and the consequent serious injury or death of manatees, has increased dramatically. As documented by the Florida Fish and Wildlife Conservation Commission, between 1979 and 2004, 1,253 manatee deaths were attributed to watercraft, out of a total of 5,033 recorded fatalities for that time period. Thus approximately one in four manatees died because of interactions with vessels. Since 2004, the percentage of total manatee mortality caused by collisions with boats has continued to hover around 25 percent, but several times the an-

nual number of watercraft-related manatee deaths has approached 100 animals. Despite decades of efforts to minimize such decimations, this threat is not under control.

Nor is the problem related simply to killing manatees. Many (and perhaps most) collisions with boats probably result in serious injuries but do not kill the struck manatee outright. Scott Calleson and Kipp Frohlich of the Office of Imperiled Species Management for the state noted that two recently necropsied manatees each had scar patterns indicative of more than fifty separate strikes by watercraft. Scientists have observed that severely injured female manatees often do not maintain the same interval between calves as healthy manatees do; instead of bearing a calf every 2.5–3 years, injured animals may double the time between calves as they devote their energy more to recovery than reproduction.

For perspective, watercraft-related deaths are consistently a significant source of annual manatee mortality, but in some years, other factors account for far more deaths. For example, 2010 and 2013 produced atypically high numbers of dead Florida manatees: 766 and 830, respectively. In 2010, exceptionally cold weather was the main culprit, accounting for the demise of at least 282 manatees (almost 37% of the total mortality). In 2013, natural causes of death, including but not limited to exposure to red tide, accounted for 197 fatalities (approximately 24% of the total).

The effects of watercraft on manatees are not limited to serious injury, impaired reproduction, and death. There is abundant evidence that manatees can detect the sounds of boats and even determine the direction from which a vessel is approaching. Given the increasing amount of boat

The parallel scars on the back of this manatee are from the blade of a boat propeller. The single vertical scar at right angles to the others is from the skeg (the back part of a boat's keel).

Manatees use their powerful flukes to swim, and sometimes part or all of a fluke is severed by a boat propeller.
It is difficult to conceive what it must be like for animals with amputated flukes, especially in how they get
around. For all their gentle and placid ways, manatees
are awfully tough animals.

traffic on finite waterways, this means that manatees are often placed
in positions where they hear and respond to boat traffic throughout the
day, and endure conditions where the underwater noise levels obstruct
communication with other manatees, including essential messages between a mother and her calf. Underwater noise levels are starting to be
viewed as a form of pollution that, in the future, needs to be mitigated as
strictly as chemical pollution levels. Imagine being in a position where
your work or other activities was interrupted many times per day by a
stimulus (visual, acoustic, or otherwise) that posed a significant and
painful threat.

Finally, watercraft in Florida are responsible for seagrass scarring and
for increased turbidity in certain locations (both of which can affect local
seagrass productivity), as well as for numerous human deaths and injuries. Impairment to the extent and productivity of seagrass beds affects
manatees, recreational fish recruitment (i.e., adding new individuals
to a population through growth, reproduction, and immigration), and
other factors of importance ecologically and to human beneficiaries.

By the time a normal, healthy female reaches age 20, she may well have produced five or six offspring. In contrast, females with one or more severe injuries may have that reproductive potential cut in half.

Despite the threat that speeding motorboats have for manatees, it is not all that unusual to see these animals nuzzling or otherwise closely approaching a propeller when the engine is turned off. The fact that manatees get struck so often is not generally an indication that boaters want to harm them; rather, manatees and boats are in close proximity through most of the Florida manatees' range, and accidental collisions result.

To help boaters avoid hitting manatees, signs are placed along Florida waterways to warn people to be observant and cautious. Additional signs inform boaters of areas where the likelihood of a collision with a manatee is sufficiently high not only to warrant slow operating speeds but also to have them enforced.

Before discussing some mitigation options, I want to be clear that the intent of the above information is *not* to rail against boats or boating. It simply represents an attempt to outline the variety of ways in which watercraft can and do represent a threat to manatees and their habitat. I enjoy boating, and the research activities of my program (e.g., photo identification) demand that my staff be on the water dozens of times each year.

Having said this, it is useful to consider the recommendations of a number of scientists and wildlife managers who advocate reductions in boat speeds as (a) a means by which manatees can detect and have time to avoid boats more easily, and (b) a way that, if a collision does occur, damage to the manatee will be minimal. This idea has governed many of the rules and regulations imposed by the responsible agencies, but, unfortunately, many boaters fail to abide by them.

In addition, Florida does not require boaters to have licenses, and until recently they did not even have to attend classes and be properly trained with regard to the rules of the road. In a regulation that finally

became effective 1 January 2010, boaters born after 1 January 1988 must possess a Florida Boating Education Safety card in order to operate a vessel powered by a motor of 10 horsepower or more.

The combination of many boats, inadequate on-water enforcement, and some inexperienced or careless operators certainly has contributed to some of the collisions vessels may have with sandbars, manatees, and other watercraft. Even some boating groups have recommended and supported mandatory education and licensing as a partial solution to this troubling problem.

## Other Sources of Disturbance to Manatees

Physical disturbances to manatees are not limited to noise or close approaches by boats. A growing issue is the presence of divers at warm springs around Crystal River in winter. Various authors have noted that when humans are present in large numbers or blatantly harass the manatees, some animals will vacate the area, despite the biological necessity of staying warm. Recent tallies suggest that somewhere between 100,000 and 125,000 people come to Crystal River each year, explicitly to have a "manatee ecotourism experience." The local community derives $20–$30 million in revenue annually from this seasonal activity and is therefore resistant to any curbs on this status quo.

The solution to this problem has been debated for many years. One option that has been recommended in a number of annual reports by the US Marine Mammal Commission, a high-ranking federal oversight agency, is for the US Fish and Wildlife Service to adopt a quota or a lottery system

What a thrill to be next to a 2,000 pound wild animal and be able to observe it closely. Such encounters can help create a corps of people who are sufficiently moved by the experience to want to engender a better conservation ethic in the general population. It remains a matter of concern, however, when hordes of people descend on one or just a few animals. Thus the thrill of an encounter must be tempered with respect for the creature's need to be unimpaired and remain a wild animal.

**A mother and her calf swim together, the latter in the classic baby position.**

that would limit the number of boats and divers that could enter the area each day, and to increase that agency's enforcement presence there. In the war of human values, however, it remains uncertain whether, and if, a thirst for short-term annual profits can be balanced with the long-term protection of animals in need of some warmth in winter.

## Climate Change

Although the effects of climate change are less well studied and understood for tropical and subtropical environments than for polar ones, it is apparent that such changes will be global. The habitats occupied by Florida manatees and other sirenians are at risk of great modification over the upcoming decades. Given the uncertainty of the nature of climate changes and of ways to mitigate them, the most precautionary step at present is to limit the effects of other, theoretically controllable issues regarding manatees, such as watercraft collisions and the availability of warm water.

Several likely scenarios associated with the effects of climate changes on Florida manatees include the following:

- *droughts*, which will reduce the availability of freshwater and freshwater habitats locally, as well as continental droughts that could magnify the spread of sub-Saharan dust, which also appears to be a factor promoting red tides (algal blooms) in the Gulf of Mexico and perhaps elsewhere
- *increased intensity and frequency of hurricanes*, which can kill a

manatee outright, as well as have significant effects on the local or regional availability of vegetation eaten by manatees

- *Increased runoff and eutrophication in local areas*, which could increase the frequency and toxicity of red tide events
- *reduction in industrial emissions* (a generally good area for mitigation efforts), which could affect the operations and longevity of power plants on which many Florida manatees rely for warmth
- *warmer waters*, in which infectious diseases may flourish and seagrasses may not
- *sediment suspension during storms*, which could lead to the increased bioavailability of sediment-bound contaminants

Altogether, we can be sure that there will be impacts of climate changes on Florida manatees and their habitat. One can only hope that the surprisingly adaptable manatees can make the necessary adjustments to survive the breadth and intensity of changes that we can only imagine and perhaps try to model.

A reduction in allowable industrial emissions may affect the operations and longevity of some power plants in Florida, especially for ones that are of greater importance as warm-water refuges for manatees than this nuclear plant at Crystal River.

## Indirect Threats to Florida Manatees

This chapter has dealt with some specific sources of danger to Florida manatees; the previous one focused on the habitat needs of this subspecies and some consequences of habitat alterations. At this point I want to shift gears and discuss some philosophical issues that can hinder the effective conservation of manatees and thereby serve as a threat to manatee well-being and survival.

The Pew Research Center conducts an annual poll of priorities by the US public. In 2015, the results of this poll indicated that environment ranks thirteenth out of the top twenty-three priorities, and global warming ranks twenty-second. Conservation, specifically, does not appear on the list. One purpose of this poll is to inform the country's leaders about what the citizens of the United States want their government to prioritize. Given the status of the US budget, it seems doubtful than any issues other than the very top few (namely, terrorism, the economy, and jobs) will get meaningful governmental support. Thus it seems uncertain, and even unlikely, that federal priorities for funding will meaningfully include conservation, the effects of global climate change, or environmental sustainability.

If one assumes, therefore, that funding for environmental and conservation issues will remain less than adequate, how can changes be made that will improve the status of those issues and the sustainability of the resources involved? Or will benign neglect and a consequent lack of funding for our environment and the living resources it contains be their downfall?

Two important conceptual tools for conservation are the precautionary approach, or precautionary principle, and burden of proof. These principles are related and can be applied to management of the risks associated with a range of threats; in this case, I apply them specifically to risks to ecosystems and species. The precautionary approach to risk management states that if an action or policy has a suspected risk of causing harm to the public or the environment, in the absence of scientific consensus that the action or policy is not harmful, the burden of proof that it is *not* harmful falls on those taking an action. For example, if someone wants to develop a coastal area that is known to include important manatee habitat but for which the exact repercussions of the development are unknown, the developer must conduct (or cause to be conducted) appropriately unbiased, independent research to demonstrate a lack of consequent harm, rather than leaving it to agencies—and thus using taxpayer funding-to show a certainty of such harm. In the absence of such definitive proof of no (or virtually no) negative effects,

the development will not be allowed, in order to be precautionary in favor of the manatees and their habitat.

Imagine the impact of establishing the precautionary principle and burden of proof as standard operating procedures. If the burden of proof resides with the entities that stand to profit from an action, rather than with taxpayer-supported agencies, the collective economic benefits for state, local, and federal governments are obvious, and a pattern would be set for making decisions *only* with adequate data to substantiate potential results that would have minimal or acceptable impacts. Although these ideas are nothing new (actually, they are decades old), the time is long past to eschew being precautionary in favor of natural resources, including manatees and other wildlife, when adequate and decisive data are lacking, and to place the burden of proof where it belongs—on the user.

These concepts are related to a more recent change in perspective associated with environmental and wildlife management: human di-

The skies at sunset over the Gulf of Mexico may be beautiful, but they make one think about what our natural world will look like as the effects of climate changes are manifested.

**Left to their own devices in the absence of human activities, manatees lead a placid and graceful life.**

mensions. In the past, and even today, conservation decisions tend to be based to a large (or even exclusive) extent on ecological or biological data. In contrast, the relatively new viewpoint of human dimensions is often lacking in approaches to informed decision making and to a broad-based understanding of the ways in which humans value, use, and depend on the natural environment. This more inclusive approach has a much higher probability of guiding decision makers toward sustainable and successful conservation decisions.

Another important need in Florida—and elsewhere—is the development and adoption of a conservation ethic, which is a shared perspective on the sustained use, allocation, and protection of resources. Conservation, properly conducted, provides a situation in which nature and human stakeholders can *both* win, but at present, conservation is sometimes portrayed as an enemy of human rights and privileges.

Without consensus on a conservation ethic, without an endorsement of the precautionary principle and an appropriate burden of proof, and without an overt embrace of a human dimensions component in decision making, the conservation of manatees and other living and nonliving resources is threatened with failure. Unless these sorts of changes in perspective occur, many future conservation decisions will cling to an unsuccessful and unacceptable status quo. In environmental impact statements, (required under prescribed circumstances by the National Environmental Policy Act, or NEPA) one option is always "no action." All too often, controversy over the other options leads inexorably to the selection of "no action," even when that option has been proven not to

This manatee is taking time for a quiet breath of air before submerging to forage.

work at present, and actually *is even more unlikely to work when the human population increases*, as is predicted in the future! At the risk of being repetitious, until and unless human population growth and its consequent impacts on global ecosystems can be successfully addressed, conservation efforts for sirenians (and other wildlife) will continue to struggle to overcome cumulative and synergistic threats, where the total effect is greater than the sum of its parts.

People in the United States and elsewhere need to change the way they do business with conservation issues, which, in turn, involves changes in their values and day-to-day activities. If people do not do so, the inertia of their current behaviors is probably the greatest single threat to manatees and other resources.

## Additional Threats to Other Sirenians

Everything stated above with regard to threats to manatees applies to some degree to other sirenian populations around the world. It is important to realize one important difference, however, between the manage-

A manatee mother and her calf swim into an uncertain future, due to human population growth, climate changes, and other threats to our environment.

ment of resources in Florida (and the United States in general), and their management in most of the countries where sirenians occupy nearby waters. People living in the vast majority of those latter countries exist at the poverty level, and conservation is not likely to be effective until such poverty is confronted, conditions improve, and people develop or acquire alternative livelihoods.

At present, poverty in many developing countries leads to ignoring protective statutes as a means of allowing communities to exploit local resources in order to acquire food and cash. As a result, sirenians are taken in gillnets and caught in other fishing gear, and they are deliberately hunted for meat, oil, and other products. The illegal bushmeat trade is not limited to the better-known hunting of terrestrial mammals, such as elephants, rhinos, and apes. For this reason, poverty-induced hunting and the incidental taking of sirenians outside of Florida is a serious conservation threat, even though these factors are rather inconsequential for Florida manatees. An important component of the solution involves the development of alternative livelihoods to replace hunting and the bushmeat trade.

Some other issues threaten sirenians in many locations outside of Florida. These include habitat and population fragmentation, due to the damming of rivers or hydroelectric power plants; the creation of "paper parks" and other protective measures that cannot be enforced and often are created with incomplete goal development and sketchy plans; and a lack of scientific information or political will with which to undergird decisions and enact changes.

Thus, whereas a number of factors threaten the conservation status and sustainability of Florida manatees, many other sirenian populations around the globe face more diverse and more serious threats. Conservation solutions must be implemented quickly and successfully, or some sirenian populations will slowly but surely be wiped out from much of their current range in the upcoming years.

Our oceans are under siege, and the tipping point for many species, where recovery becomes impossible, may be near.

# 9

# *Conservation Solutions*

Conservation, I believe, can and should be thought of as having four parts: (1) it should be conducted in a manner in which multiple human stakeholders and the environment can both win, (2) it must allow enough time and encourage efforts to develop conservation strategies that lead to optimal outcomes, (3) its advocates must not be daunted by the enormity of the challenges and a lack of vision among the world's leaders with regard to what conservation can and should achieve now and in the future, and (4) if current generations do not change the way conservation is valued and undertaken, we will leave the world a much poorer place in terms of its biodiversity, aesthetics, and sustainability. For instance, our oceans are under siege. According to marine biologist Sylvia Earle, 90 percent of all large sharks, tuna, marlin, and sailfish are gone, as well as half of the world's coral reefs, seagrass beds, and mangroves.

Having said all that, this is undoubtedly the most important chapter of the book. Its goals include defining the need for improvements in conservation and a new way of doing business; describing the components of effective conservation; underscoring the role of science in conservation, as well as understanding the fundamental differences between the two; reviewing conservation tools; and recommending organizational and personal approaches and attributes for success.

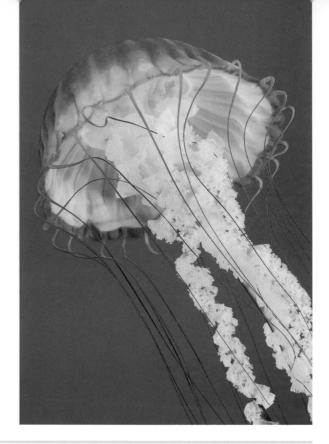

Sea nettles may be beautiful, but in recent decades, more-frequent blooms of these and other jellyfish signal the degraded state of our oceans. Jellyfish foul fishing nets and plug intake pipes at power plants and desalination plants. Moreover, their stings do nothing to enhance diving or swimming.

This chapter had its roots in many conversations with valued colleagues including (but certainly not limited to) Helene Marsh, Tom O'Shea, Benjamin Morales, Tim Ragen, Pat Rose, David Laist, and the late John Twiss. In a clarion call to action, the 2003 report by the Pew Oceans Commission stated that the "oceans are in crisis and the stakes could not be higher.... Without reform, our daily actions will increasingly jeopardize a valuable natural resource." At approximately the same time, international efforts arose to stem the loss of biodiversity; specifically, the Convention on Biological Diversity emerged in the latter part of the twentieth century to address increased global biodiversity loss, the sustainable use of resources, and a fairer and more equitable sharing of the benefits of genetic resources. This convention selected 2010 (seemingly, a very long time ago) as the International Year of Biodiversity and designated 2011–2020 as a timeframe for the development of a Strategic Plan for Biodiversity.

Sadly, despite the dedicated efforts of many individuals, global biodiversity is anything but assured. Worldwide, there are approximately 64,000 species that have undergone population assessments, and one in three of them (nearly 20,000 species) is threatened with extinction. Perhaps even more startling is the reality that the vast majority of the estimated 8.7 million species on Earth today (of which only 1.2 million have been catalogued) have not been assessed, because humans tend to focus on those that we deem "important," typically for economic reasons. If people are so evidently failing to conserve the species on which we

place some value, what are we likely to be doing with efforts to conserve those species that we do not prize as much?

Marine mammals are among the world's most charismatic species. Surely species we greatly treasure should be doing well, but 2.5 percent (3 out of ~120 species*) of the marine mammal species have become extinct in roughly the last sixty years. Several species (e.g., northern elephant seals [*Mirounga angustirostris*] and gray whales [*Eschrichtius robustus*] in the eastern North Pacific Ocean) have experienced remarkable recoveries but still may be unexpectedly vulnerable, due to a limited gene pool and inbreeding during that recovery. Moreover, some marine mammal taxa remain in critical condition and may not persist much longer at all. These include the vaquita (*Phocoena sinus*, also called the Gulf of California harbor porpoise), the Mediterranean monk seal (*Monachus monachus*), and the AT1 pod of killer whales (*Orcinus orca*) in Alaska.

Among the various sirenian species, all are threatened with extinction, but the West African manatee seems especially vulnerable. The Antillean subspecies will probably disappear from parts of its range in the near future, even though a few populations may persist for some time.

In a 2007 assessment of the status of endangered marine mammals in the United States, the Marine Mammal Commission (the federal agency with oversight for marine mammal research and conservation in

In the late 1800s, following decades of hunting, there were fewer than 100 northern elephant seals left. They are actually one of the few marine mammal species to have recovered from overharvest, and there are currently more than 150,000 of these gigantic seals off the west coast of North America. The dramatic reduction in their population size, however, led to reduced genetic diversity (i.e., a genetic bottleneck), which leaves the northern elephant seal species vulnerable to this day.

*The marine mammals that have become extinct in recent decades are the baiji (Yangtze river dolphin), Caribbean monk seal, and Japanese sea lion.

In contrast to the situation with northern elephant seals, numbers for Hawaiian monk seals are low and continue to decline. Even with only 1,000 or so remaining individuals, Hawaiian monk seals are still more numerous than their close relative, Mediterranean monk seals. A third species, the Caribbean monk seal, was wiped out by the 1950s.

the United States) indicated that twenty-two taxa of marine mammals in US waters are endangered, threatened, or depleted. Of these, one (the Caribbean monk seal, *Monachus tropicalis*) is actually extinct, one (the AT1 pod of killer whales) has not bred successfully for thirty years and thus is not viable (capable of surviving), and five are estimated to have fewer than 400 individuals remaining. If a grade for marine mammal conservation was being provided on a report card, it would be an F, despite the herculean efforts of some individuals and groups. Helene Marsh, Tim Ragen, and I (among others) have concluded that unless fundamental changes occur in the way we do business, the future for marine mammals looks bleak.

How can this be possible? People *do* care about manatees and other marine mammals, so how can these animals be slipping away from us? And what does this suggest about the status of species that humans value less? A lack of scientific information is often mentioned as the primary impediment to species conservation, but that is not the case. Science does, indeed, play a crucial role in conservation, but for many species, including the West Indian manatee, scientific research and data are sufficient to understand and mitigate most threats *if political will to do so exists*. Science and conservation, albeit interconnected in some ways, are fundamentally different. Science is designed to test hypotheses and answer questions, whereas the purpose of conservation is to solve problems. Conservation works best when decisions are supported by good science, but it is important to understand that conservation requires

A curious manatee leaves the group for a better look at the photographer.

more than just solid data, and conservation strategies must achieve their objectives even in the face of scientific uncertainty.

Thus a vital consideration for scientists interested in positively affecting conservation is not, "What new studies or monitoring can I acquire the funding to do?" Rather, the most critical question for scientists intent on promoting conservation is, "How can we communicate our results in a way that is compelling to decision makers and that ultimately speaks to human values in a way that affects human behaviors?" The issue is not a lack of science as much as it is the extent of what societies are willing to do—based on available science *and human values*—to protect the future. In short, the primary impediment to species conservation is an unwillingness to conserve, which relates to values more than to data.

Until now, this and previous chapters have used the word "conservation" without definition. Although people have characterized the term in a number of ways, I prefer the following delineation from Gary Meffe and

Scientists sometimes attach satellite tags to manatees, in order to learn where the animals go and the specific locations they prefer. Such information aids decisions by managers who seek to protect the places that manatees need and use. The tags do not impair the movements of manatees, and a weak link in the "tether" is designed to break if the tether becomes entangled.

his coauthors: "Conservation seeks an ideal relationship between humans and the natural world. That relationship safeguards the viability of all biota and the ecosystems on which they depend, while allowing human benefit, for present and future generations, through various consumptive and non-consumptive uses." The especially pertinent attributes of this definition (or description) include an extension to ecosystems as well as to species of interest; a recognition of benefits for the range of human stakeholders; a focus on the present and, more importantly, the future; and the acknowledgment and acceptance of certain consumptive uses.

In an idealized world, it would be nice to have exhaustive data and information on species and ecosystems of concern before implementing conservation strategies. In the real world, this is generally impractical, and adaptive management (the acquisition of and adjustments to new findings) needs to occur in parallel with solving problems based on the best available information. The following example illustrates the need for conservation strategies to solve problems (or stop the bleeding, to use a medical analogy), rather than to monitor the demise (or document

Manatees are routinely captured by experienced teams of scientists, in order to perform health assessments or attach tracking tags. The US Geological Survey has been conducting such captures since the 1970s; the group pictured here is working at Crystal River.

Most species of large sharks, such as this great white shark swimming off Guadalupe Island in Baja California, have been overharvested. In many cases, only the fin is removed, destined for markets in Asia. Although this cruel and wasteful activity continues to occur, many markets for shark fins have disappeared, in response to a global outcry.

the rate of blood loss) of a dugong population being perilously impacted by human activities.

About a decade ago, several scientists and conservationists (myself included) were invited to come to Mozambique to consider how best to try to conserve the largest remaining group of dugongs in East Africa. These dugongs, numbering just a couple of hundred individuals, occupy the waters of Bazaruto Archipelago National Park in northern Mozambique. There, despite legal protection by the government, dugongs were being taken by people in local subsistence communities, but these animals were not the primary target of the fishery. Rather, Asian-based markets were paying up to US$100 per kilo (2.2 pounds) for shark fins, and the

WORLD WILDLIFE FUND FIRST DAY COVER

5/
KENYA
ENDANGERED SPECIES    DUGONG    NGUVA

26 SP 77 KENYA

FIRST DAY OF ISSUE
OFFICIAL COMMEMORATIVE STAMP OF THE REPUBLIC OF KENYA

Dugongs in Kenya have become extremely rare, due to overharvesting. Nonetheless, the dugong remains iconic in that country and is celebrated on this postage stamp.

gillnets employed to catch sharks incidentally captured dugongs, sea turtles, and inshore dolphins, all of which were eaten or traded.

Complicating the situation was the fact that Mozambique is one of the world's poorest countries. In 2012, its economic status, plus other factors, positioned the country in 185th place out of 187 ranked countries and territories on the Human Development Index; for comparison, the United States is typically among the top-ranked countries in the world on this index. When the dugong workshop was taking place, the average *annual* income per family in Mozambique was in the neighborhood of US$200. The capture of a single large shark could double a family's yearly income, and if meat and other products were acquired as a consequence of their shark-fishing efforts, the community could benefit enormously. Despite laws designed to protect dugongs and other resources, poverty, poor health, and inadequate nutrition compelled coastal subsistence communities in Mozambique to ignore the laws, in an effort to survive. The point is that in developing countries, it is especially difficult to achieve conservation goals without addressing poverty and offering alternative livelihoods.

At the dugong workshop, the government of Mozambique was initially most interested in conducting well-designed aerial surveys to document population trends for the Bazaruto dugongs. While the participating scientists did not say that such surveys were without value (although it is difficult and relatively expensive to design and conduct surveys that can statistically document population sizes or trends with

accuracy or precision for hard-to-spot marine mammal populations), they emphatically advised the government that the highest priority was to solve the problem of dugong captures associated with the gillnet fishery for sharks. In other words, it was much more important to stop the bleeding than to document rate of blood loss. Fortunately, that advice was heeded, and even though dugong population surveys have been done, there have also been productive efforts to empower and involve community members in alternative jobs, including serving as wildlife or ecotourism guides and local rangers. The Bazaruto dugongs are not out of danger, but Mozambique's parallel efforts to address poverty and encouraging different livelihoods while promoting a conservation ethic and augmenting scientific data continues.

West Indian manatees have a broad range, which includes the waters of nearly two dozen nations, most of which are developing countries where (a) manatees are reduced in numbers, (b) habitat for manatees is compromised, and (c) people occasionally kill manatees (deliberately

West Indian manatees can be found in coastal and riverine habitats of nearly two dozen countries or territories, ranging from the United States to Brazil, including a number of Caribbean islands.

Bob Bonde of the US Geological Survey has worked with manatees for nearly four decades. Given his experience and expertise, he often leads efforts to capture and evaluate West Indian manatees in Florida and elsewhere in the species' range.

or incidentally) and use them for food. For West African and Amazonian manatees, as well as many regional dugong populations, the same factors hold true. The example provided by the Bazaruto dugongs is thus highly appropriate and germane on a regional basis for other sirenians.

It would certainly be useful to establish a general blueprint for the conservation of manatees and dugongs that could be adapted and applied to particular locations and situations. Such a blueprint for manatees and dugongs has been articulated by Helene Marsh, Tom O'Shea, and me, and it includes the following components (emphasis added):

- establish clear *goals*
- recognize that conservation is *value based*
- adopt a geocentric (worldwide) conservation *ethic*
- recognize the central role of *values* in formulating public policy
- establish fundamental *principles* for twenty-first-century conservation
- identify problems (or threats) and seek *solutions*
- clarify and address issues of *scale*
- establish *long-term* funding for programs
- develop a coherent *ideology*
- frame the debate *in an anticipatory manner*
- provide inspired and inspiring *leadership*
- build *infrastructure*
- be *proactive* (focused on the future), not reactive (responding to the past)

Florida-wide, the official state marine mammal appears on thousands of license plates.Crystal River, Florida, is arguably the most recognized place to see manatees in the world. The city has publically embraced its manatees, with representations of the species used on its water tower, as mailboxes, and at City Hall.

- be *creative*, not tied to traditional approaches
- build *interdisciplinary* teams
- err on the side of too much *communication*, not too little

This blueprint obviously covers a range of factors, but especially critical ones to integrate into a plan and use as the basis for effective conservation include recognizing the importance of *values*, ensuring that the *worth of conservation* is clearly an essential part of the whole, being *proactive*, nurturing inspired and inspirational *leadership*, and seeking *solutions*, not just more information.

Marsh, O'Shea, and I also discuss tools that can help manatee (and other) conservation efforts succeed. Those tools fall into two general categories: regulatory tools and enabling ones. Regulatory tools currently exist globally. They include legal protection for manatees, the enforcement of laws, and the creation of aquatic protected areas. Even though these measures are rather common and routine, they are frequently ineffective in achieving conservation goals and objectives. Part of the

A manatee peacefully surfaces to breathe. Despite many threats to their existence, greater awareness of successful approaches to the conservation both of this species and of the animals' habitat offers hope that manatees will persist forever. The outcome depends on human values and will.

problem is that regulatory tools are not designed in a way that promotes their effectiveness; for example, many aquatic protected areas are "paper parks" that exist in documents and on maps but fail to achieve meaningful conservation outcomes. In addition, whereas regulatory tools are not without value when properly designed, implemented, and enforced, by themselves they are not likely to achieve the community buy-in and compliance necessary for success. For example, many drivers exceed the posted speed limits on roadways, which illustrates how laws alone do not necessarily compel responsible human actions.

Enabling tools warrant particular attention to help achieve the goals listed in the conservation blueprint. Specifically, the enabling tools that have been shown to work with manatees and many other species include the following:

- education and awareness activities
- development of community partnerships
- creation of cross-species initiatives
- use of charismatic or economically important species as flagship or umbrella species, since effective conservation strategies will also protect myriad other species
- reinforcement of cultural protocols
- scientific research (which is still important but is not the whole solution)
- management for multiple threats
- use of spatial management to assess risks (and options)

- incorporation of adaptive management
- consideration of a range of novel economic incentives and other financial tools

Several of these enabling tools are self-explanatory or have been referred to in this or other chapters. Two enabling strategies that seem to warrant additional explanation, however, are the use of spatial management and economic considerations .

Helene Marsh and her current and former students at James Cook University in northern Queensland have championed the use of spatial tools for dugong management in the Great Barrier Reef and other multi-use locations in Australia. Several individuals, including Ken Haddad and Richard Flamm of the Florida Fish and Wildlife Conservation Commission, have done the same for Florida manatees. Basically, spatial analysis provides a tool that shows multiple, geographically overlapping data-bases, such as species distributions and densities, areas of high human-related mortality, and locations where habitat has been compromised. If a new activity is proposed, data are presented as an image, using overlying databases to demonstrate points of either potential conflict or no conflict. This can help managers minimize or avoid unwanted and unnecessary harm to sirenians. Spatial analysis using geographic information systems is an outstanding tool for communicating with decision makers.

Economic enabling tools allow conservationists to fight fire with fire. In our current value system, few things matter more to people than making money. Yet if it can be shown that conservation, done properly, financially profits a range of human stakeholders, then conservation de facto becomes a high priority. One such tool is an environmental mort-gage, a form of microfinancing in which villagers or other local entities, after abiding by a set of rules that protect local natural resources, acquire a fund that can be used for schools, community services, and basic health care. In the developing world in particular, environmental mortgages create a win-win situation, providing tools and services that people in subsistence communities or in countries in the bottom segment of the Human Development Index need, without compromising the natural environment on which such communities or countries rely for their future. Other economic tools and incentives also address the issues of poverty and alternative livelihoods in ways that allow local communities to enhance their well-being without damaging their future by destroying resources.

In developing countries, it is especially difficult to put conservation measures into practice without also confronting the issue of poverty. One solution is to provide livelihoods that do not depend on harvesting manatees and other wild species. A recent project in Nigeria offers an elegant, repeatable approach for reducing or even eliminating harvests of West African manatees. In the community of Owode-Ise, located in Lagos State on Lekki Lagoon, manatees have been hunted using wooden trigger-traps. The meat and other products have been used by the community, as well as traded and sold. In 2012, eleven hunters from Owode-Ise killed 132 manatees. In 2013, Nigerian agency officials responded by offering the hunters an alternative livelihood, catfish (*Clarias gariepinus*) aquaculture, in exchange for removing their traps and ceasing to kill manatees. The former hunters were provided with training, as well as all necessary materials for catfish aquaculture. As a result, in 2013–2014, no manatees were killed in Owode-Ise, and the former hunters harvested more than 2,645 pounds (1,200 kilograms) of catfish in their first year of operation. The community realized that over the long term, catfish are a more reliable and larger source of income than manatee hunting. Word of their success has spread, and three more communities located on Lekki Lagoon have requested aquaculture training in exchange for stopping their manatee hunting. The broader implications of this project are enormous in terms of effective and sustained protection for West African manatees.

The use of science to understand manatee biology and ecology is not an easy endeavor, but the careful administration of conservation measures for manatees, supported by good science and inclusive of multiple stakeholders, is even harder to achieve. Unless society uses a collection of enabling tools, redefines and reprioritizes values, and embraces the new way of doing business exhorted by the Pew Oceans Commission, it seems unlikely that upcoming generations will succeed in conserving manatees and other living resources any better than recent ones have. With the astronomical growth of the global human population, estimated to eclipse 9 billion by the year 2050 (an increase of almost 30% over the next forty years), conservation will be much more difficult to achieve in the future, unless strategies and values change.

At the organizational level, some general perspectives may be helpful. Conservation advocates should recognize that multiple value systems exist, but they should also strive to have society place a very high value on conservation. It will be important to include multiple stakeholders in order to reach solutions; leaving groups out of the discussion does nothing to make the excluded parties want to interact in good faith. Conservation leaders must be honest, open, and proactive, and they should insist that the best available information be used in making decisions. Perhaps most importantly, conservation advocates must not wait to act until all possibilities are known. They should recognize what the obstacles to conservation are and promote society's will to confront and address those obstacles and bring about change.

The Florida Everglades are hauntingly beautiful. Even more important for manatees is the fact that this vast area provides wonderful habitat, with relatively little human intrusion. The protected waters of Everglades National Park represent a long-term stronghold for the manatee population in Florida.

Sustaining Florida manatees into the future is not assured, but the responsible agencies and the citizens of Florida deserve a lot of credit for actions and attitudes that have led to a notable increase in the size of the state's manatee population.

At the individual level, conservation will provide a stern test of character. If you want to become engaged, you must become proactive. You must have the guts to stand up for conservation principles and values. You must be creative, opportunistic, and imaginative. You must address and help shape human behaviors and what people consider worthwhile, with a focus on winning wars, not every battle. You must learn to work well with others whose strengths and perspectives differ from your own. You will be tested to use finite human and financial resources in ways that make a positive mark. And, most importantly, you must focus on solutions.

Conservation champions do not stand alone, and role models from whom we can gain strength and inspiration abound. Mine is a great lady, Marjory Stoneman Douglas (1890–1998), who published *The Everglades: River of Grass* in 1947, which facilitated the establishment of Everglades National Park. In the face of immense pressure to relegate all of southern Florida to farms, Douglas had the knowledge, dedication, passion, communication skills, toughness, and set of values to make a difference in both the quality of life for Floridians and the sustainability of Florida's manatees and ecosystems. Many readers of the present book also possess those traits, so with proper dedication, and perhaps a little luck, they may someday be ranked as conservation leaders, alongside Marjory Stoneman Douglas.

# *Epilogue*

On 8 January 2016, the US Fish and Wildlife Service published a proposal indicating that the West Indian manatee (including both subspecies) had recovered sufficiently to warrant downlisting from an endangered to a threatened species under the US Endangered Species Act (ESA). This change in status would not mandate any reductions in federal protection of manatees, although state and other agencies may or may not maintain legal regulations under their jurisdictions. Practically, downlisting has the potential, under pressure from some stakeholder groups, to reduce protective measures for manatees. The Fish and Wildlife Service has therefore solicited public comments to guide its final decision.

This downlisting proposal is not unexpected, since the number of Florida manatees has grown in recent years, due to protective measures and compliance with them. Yet in the face of ongoing habitat degradation, poorly controlled threats to manatees, and even deaths from uncertain causes, an important question still exists. Does this decision reflect an adequately precautionary approach to conservation? If the human population in Florida grows by 30 percent in the next fifteen years, as expected, can consequent effects on manatee habitat and on the animals themselves be controlled and mitigated? In addition, if protection under the ESA actually is reduced, and if the manatee population declines, cur-

rent survey methods are imprecise enough for their reduced numbers not to be spotted for some time, suggesting that drops in population and species levels could go undetected and unmitigated for a considerable period.

For the Antillean subspecies, populations of manatees are either declining or of uncertain status in 84 percent of the nations in whose waters they are found, and threats to these animals are poorly controlled. In some countries, this means that there probably will be fewer than 100 manatees. Thus manatees could disappear from the majority of their Caribbean range in the not-too-distant future.

The people of Florida, in particular, deserve great credit for adjusting their behaviors in ways that have allowed that state's manatee population to rebound from decades, and even centuries, of excessive takes. The US Fish and Wildlife Service, as the responsible federal agency, has indicated that the downlisting of this species would not be accompanied by any diminution of protection for it. That is excellent news, but, as I believe and have noted in my comments to the US Fish and Wildlife Service, until this agency conducts more-complete analyses of existing data to ensure that the best-available information is used in making its final decision, and until threats to manatees are better understood and under control now and in the future, it is premature to downlist the species. An important preliminary step, in my opinion, is that before any change in status is adopted, the agencies should work with communities throughout the range of the West Indian manatee to adopt a strong conservation ethic and thus continue to ensure this species' future.

# Selected References and Further Reading

**General**

Glaser, K., and J. E. Reynolds III. 2003. Mysterious Manatees. University Press of Florida, Gainesville. 187 pp.

Hines, E. M., J. E. Reynolds III, L. V. Aragones, A. A. Mignucci-Giannoni, and M. Marmontel (eds.). 2012. Sirenian Conservation: Issues and Strategies in Developing Countries. University Press of Florida, Gainesville. 326 pp.

Marsh, H., T. J. O'Shea, and J. E. Reynolds III. 2011. Ecology and Conservation of the Sirenia: Dugongs and Manatees. Cambridge University Press, Cambridge. 521 pp.

Reep, R. L., and R. K. Bonde. 2006. The Florida Manatee: Biology and Conservation. University Press of Florida, Gainesville. 189 pp.

Reynolds, J. E., III, and D. K. Odell. 1991. Manatees and Dugongs. Facts on File, New York. 192 pp.

**Chapter 1**

Reynolds, J., J. Provancha, L. Morris, K. Scolardi, and J. Gless. 2015. The Indian River Lagoon: The world's most important habitat to West Indian manatees is enormously vulnerable and in urgent need of protection. Paper presented at 21st Biennial Conference on Marine Mammals, 13–18 December 2015, San Francisco, CA.

United Nations Environment Programme (UNEP). 2010. Regional Management Plan for the West Indian Manatee (*Trichechus manatus*), compiled by E. Quintana-Rizzo and J. E. Reynolds III. CEP Technical Report No. 48. UNEP, Caribbean Environment Programme (CEP), Kingston, Jamaica. 124 pp.

**Chapter 2**

Elsner, R. 1999. Living in water: Solutions to physiological problems. Pp. 73–116. IN: J. E. Reynolds III and S. A. Rommel (eds.). Biology of Marine Mammals. Smithsonian Institution Press, Washington, DC. 578 pp.

Marsh, H., T. J. O'Shea, and J. E. Reynolds III. 2011. Ecology and Conservation of the Sirenia: Dugongs and Manatees. Cambridge University Press, Cambridge. 521 pp.

Marshall, C. D., G. D. Huth, V. M. Edmonds, D. L. Halin, and R. L. Reep. 1998. Prehensile use of perioral bristles during feeding and associated behaviors of the Florida manatee (*Trichechus manatus latirostris*). Marine Mammal Science 14:274–289.

Pabst, D. A., S. A. Rommel, and W. A. McLellan. 1999. The functional morphology of marine mammals. Pp. 15–72. IN: J. E. Reynolds III and S. A. Rommel (eds.). Biology of Marine Mammals. Smithsonian Institution Press, Washington, DC. 578 pp.

Reep, R. L., and R. K. Bonde. 2006. The Florida Manatee: Biology and Conservation. University Press of Florida, Gainesville. 189 pp.

Reynolds, J. E., III, and C. D. Marshall. 2012. Vulnerability of sirenians. Pp. 12–19. IN: E. M. Hines, J. E. Reynolds III, A. A. Mignucci-Giannoni, L. V. Aragones, and M. Marmontel (eds.). Sirenian Conservation: Issues and Strategies in Developing Countries. University Press of Florida, Gainesville. 326 pp.

Reynolds, J. E., III, D. K. Odell, and S. A. Rommel. 1999. Marine mammals of the world. Pp. 1–14. IN: J. E. Reynolds III and S. A. Rommel (eds.). Biology of Marine Mammals. Smithsonian Institution Press, Washington, DC. 578 pp.

Reynolds, J. E., III, and S. A. Rommel. 1996. Structure and function of the gastrointestinal tract of the Florida manatee, *Trichechus manatus latirostris*. Anatomical Record 245(3):539–558.

Ricklefs, R. E. 1990. Ecology, 3rd edition. W. H. Freeman, New York. 896 pp.

Rommel, S. A., and J. E. Reynolds III. 2000. Diaphragm structure and function in the Florida manatee (*Trichechus manatus latirostris*). Anatomical Record 259(1):41–51.

Rommel, S. A., J. E. Reynolds III, and H. A. Lynch. 2003. Adaptations of the herbivorous marine mammals. Pp. 287–308. IN: L. 't. Mannetje, L. Ramirez-Aviles, C. Sandoval-Castro, and J. C. Ku-Vera (eds.). Matching Herbivore Nutrition to Ecosystems Biodiversity: VI International Symposium on the Nutrition of Herbivores; Proceedings of an International Symposium held in Mérida, Mexico, 19–24 October 2003, Universidad Autónoma de Yucatán. 341 pp.

Rose, P. M., and T. Stubbs (directors). 1981. Silent Sirens: Manatees in Peril. Film and VHS. Produced by Florida Audubon Society and Save the Manatee Club, Maitland, FL.

## Chapter 3

Domning, D. P. 2001. Sirenians, seagrasses, and Cenozoic ecological change in the Caribbean. Paleogeography, Paleoclimatology, Paleoecology 166:27–50.

Marsh, H., T. J. O'Shea, and J. E. Reynolds III. 2011. Ecology and Conservation of the Sirenia: Dugongs and Manatees. Cambridge University Press, Cambridge. 521 pp.

## Chapter 4

Domning, D. P. 1982. Commercial exploitation of manatees *Trichechus* in Brazil c. 1785–1973. Biological Conservation 22:101–126.

Domning, D. P. 1991. Why save the manatee? Pp. 167–174. IN: J. E. Reynolds III and D. K. Odell. Manatees and Dugongs. Facts on File, New York. 192 pp.

Glaser, K., and J. E. Reynolds III. 2003. Mysterious Manatees. University Press of Florida, Gainesville. 187 pp.

O'Shea, T. J. 1988. The past, present, and future of manatees in the southeastern United States: Realities, misunderstandings, and enigmas. Pp. 184–204. IN: R. R. Odum, K. A. Riddleberger, and J. C. Ozier (eds.). Proceedings of the Third Southeastern Nongame and Endangered Wildlife Symposium. Georgia Department of Natural Resources, Game and Fish Division, Social Circle. 253 pp.

Reynolds, J. E., III, and D. K. Odell. 1991. Manatees and Dugongs. Facts on File, New York. 192 pp.

Reynolds, J. E., III, and R. S. Wells. 2003. Dolphins, Whales, and Manatees of Florida: A Guide to Sharing Their World. University Press of Florida, Gainesville. 148 pp.

## Chapter 5

Deutsch, C. J., J. P. Reid, R. K. Bonde, D. E. Easton, H. I. Kochman, and T. J. O'Shea. 2003. Seasonal movements, migratory behavior, and site fidelity of West Indian manatees along the Atlantic Coast of the United States. Wildlife Monographs 151:1–77.

Hartman, D. S. 1979. Ecology and Behavior of the Manatee (*Trichechus manatus*) in Florida. Special Publication No. 5. American Society of Mammalogists, [Pittsburgh]. 153 pp.

Marsh, H., T. J. O'Shea, and J. E. Reynolds III. 2011. Ecology and Conservation of the Sirenia: Dugongs and Manatees. Cambridge University Press, Cambridge. 521 pp.

Ricklefs, R. E. 1990. Ecology, 3rd edition. W. H. Freeman, New York. 896 pp.

Wells, R. S., D. J. Boness, and G. B. Rathbun. 1999. Behavior. Pp. 324–422. IN: J. E. Reynolds III and S. A. Rommel (eds.). Biology of Marine Mammals. Smithsonian Institution Press, Washington, DC. 578 pp.

## Chapter 6

Boyd, I. L., C. Lockyer, and H. Marsh. 1999. Reproduction in marine mammals. Pp. 218–286. IN: J. E. Reynolds III and S. A. Rommel (eds.). Biology of Marine Mammals. Smithsonian Institution Press, Washington, DC. 578 pp.

Marsh, H., T. J. O'Shea, and J. E. Reynolds III. 2011. Ecology and Conservation of the Sirenia: Dugongs and Manatees. Cambridge University Press, Cambridge. 521 pp.

Runge, M. C., C. A. Langtimm, and W. L. Kendall. 2004. A stage-based model of manatee population dynamics. Marine Mammal Science 20:361–385.

**Chapter 7**

Ackerman, B. B., S. D. Wright, R. K. Bonde, D. K. Odell, and D. J. Banowetz. 1995. Trends and patterns of mortality of manatees in Florida, 1974–1992. Pp. 223–258. IN: T. J. O'Shea, B. B. Ackerman, and H. F. Percival (eds.). Population Biology of the Florida Manatee. US Department of the Interior, National Biological Service, Washington, DC. 289 pp.

Bossart, G. D., R. A. Meisner, S. A. Rommel, S. Ghim, and A. B. Jensen. 2002. Pathological features of the Florida manatee cold stress syndrome. Aquatic Mammals 29:9–17.

Flamm, R. O., J. E. Reynolds III, and C. Harnak. 2012. Mapping and characterizing the network of warm-water sites used by manatees (*Trichechus manatus latirostris*) along the southwestern coast of Florida. Environmental Management 51(1):154–166.

Gannon, J. G., K. M. Scolardi, J. E. Reynolds III, J. K. Koelsch, and T. J. Kessenich. 2007. Habitat selection by manatees in Sarasota Bay, Florida. Marine Mammal Science 23(1):133–143.

Hartman, D. S. 1979. Ecology and Behavior of the Manatee (*Trichechus manatus*) in Florida. Special Publication No. 5. American Society of Mammalogists, [Pittsburgh]. 153 pp.

Marsh, H., T. J. O'Shea, and J. E. Reynolds III. 2011. Ecology and Conservation of the Sirenia: Dugongs and Manatees. Cambridge University Press, Cambridge. 521 pp.

Miksis-Olds, J. L., P. L. Donaghay, J. H. Miller, P. L. Tyack, and J. A. Nystuen. 2007. Noise level correlates with manatee use of foraging habitats. Journal of the Acoustical Society of America 121:3011–3020.

Ortiz, R. M., G. A. J. Worthy, and D. S. MacKenzie. 1998. Osmoregulation in wild and captive West Indian manatees (*Trichechus manatus*). Physiological Zoology 71:449–457.

O'Shea, T. J., C. A. Beck, R. K. Bonde, H. I. Kochman, and D. K. Odell. 1985. An analysis of manatee mortality patterns in Florida, 1976–81. Journal of Wildlife Management 49:1–11.

**Chapter 8**

Calleson, C. S., and R. K. Frohlich. 2007. Slower boat speeds reduce risks to manatees. Endangered Species Research 3:295–304.

Hines, E. M., J. E. Reynolds III, L. V. Aragones, A. A. Mignucci-Giannoni, and M. Marmontel. 2012. Sirenian Conservation: Issues and Strategies in Developing Countries. University Press of Florida, Gainesville. 326 pp.

Marine Mammal Commission. 2007. The Biological Viability of the Most Endangered Marine Mammals and the Cost-Effectiveness of Protection Programs: A Report to Congress by the Marine Mammal Commission. US Marine Mammal Commission, Bethesda, MD. 448 pp.

Marsh, H., T. J. O'Shea, and J. E. Reynolds III. 2011. Ecology and Conservation of the Sirenia: Dugongs and Manatees. Cambridge University Press, Cambridge. 521 pp.

Runge, M. C., C. A. Sanders-Reed, and C. J. Fonnesbeck. 2007. A core stochastic population projection model for Florida manatees (*Trichechus manatus latirostris*). US Geological Survey Open-File Report 2007-1082. 41 pp.

Runge, M. C., C. A. Sanders-Reed, C. A. Langtimm, and C. J. Fonnesbeck. 2007. A quantitative threats analysis for the Florida manatee (*Trichechus manatus latirostris*). US Geological Survey Open-File Report 2007-1086. 34 pp.

US Fish and Wildlife Service. 2001. Florida Manatee Recovery Plan (*Trichechus manatus latirostris*), 3rd revision. US Fish and Wildlife Service, Atlanta. 144 pp. + appendices.

Wetzel, D. L., E. Pulster, and J. E. Reynolds III. 2012. Organic contaminants and sirenians. Pp. 196–203. IN: E. M. Hines, J. E. Reynolds III, A. A. Mignucci-Giannoni, L. V. Aragones, and M. Marmontel (eds.). Sirenian Conservation: Issues and Strategies in Developing Countries. University Press of Florida, Gainesville. 326 pp.

Wetzel, D. L., J. E. Reynolds III, J. M. Sprinkel, L. Schwacke, P. Mercurio, and S. A. Rommel. 2010. Fatty acid signature analysis as a potential forensic tool for Florida manatees (*Trichechus manatus latirostris*). Science of the Total Environment 408:6124–6133.

## Chapter 9

Bolaji, D., O. Fakayode, O. Aliu, I. Ayaobu-Cookey, A. Akintayo, S. Opurum, T. Diagne, D. Bolaji, and L. Keith-Diagne. 2015. Conservation of the African manatee (*Trichechus senegalensis*) through catfish aquaculture as an alternative livelihood for hunters in Owode-Ise, Lagos State, Nigeria. Paper presented at 21st Biennial Conference on Marine Mammals, 13–18 December 2015, San Francisco, CA.

Deutsch, C. J., and J. E. Reynolds III. 2012. Florida manatee status and conservation issues: A primer. Pp. 23–35. IN: E. M. Hines, J. E. Reynolds III, A. A. Mignucci-Giannoni, L. V. Aragones, and M. Marmontel (eds.). Sirenian Conservation: Issues and Strategies in Developing Countries. University Press of Florida, Gainesville. 326 pp.

Douglas, M. S. 1947. The Everglades: River of Grass. Rinehart, New York. 406 pp.

Hines, E. M., J. E. Reynolds III, L. V. Aragones, A. A. Mignucci-Giannoni, and M. Marmontel. 2012. Sirenian Conservation: Issues and Strategies in Developing Countries. University Press of Florida, Gainesville. 326 pp.

Marsh, H., P. Arnold, M. Freeman, D. Haynes, D. Laist, A. Read, J. Reynolds III, and T. Kasuya. 2003. Strategies for conserving marine mammals. Pp. 1–19. IN: N. Gales, M. Hindell, and R. Kirkwood (eds.). Marine Mammals: Fisheries, Tourism, and Management Issues. CSIRO Publications, Collingwood, Victoria, Australia, 446 pp.

Marsh, H., and B. Morales-Vela. 2012. Guidelines for developing protected areas for sirenians. Pp. 228–234. IN: E. M. Hines, J. E. Reynolds III, A. A. Mignucci-Giannoni, L. V. Aragones, and M. Marmontel (eds.). Sirenian Conservation: Issues and Strategies in Developing Countries. University Press of Florida, Gainesville. 326 pp.

Marsh, H., T. J. O'Shea, and J. E. Reynolds III. 2011. Ecology and Conservation of the Sirenia: Dugongs and Manatees. Cambridge University Press, Cambridge. 521 pp.

Meffe, G. K., W. F. Perrin, and P. K. Dayton. 1999. Marine mammal conservation: Guiding principles and their implementation. Pp. 437–454. IN: J. R. Twiss Jr. and R. R. Reeves (eds.). Conservation and Management of Marine Mammals. Smithsonian Institution Press, Washington, DC. 471 pp.

Pew Oceans Commission. 2003. America's Living Oceans: Charting a Course for Sea Change. www.pewtrusts.org/~/media/assets/2003/06/02/full_report.pdf

Reynolds, J. E., III, H. Marsh, and T. J. Ragen. 2009. Marine mammal conservation. Journal of Endangered Species Research 7(1):23–28.

## Selected Websites for Further Information

Convention on International Trade in Endangered Species of Wild Flora and Fauna: www.cites.org.

Federal Register (for access to US Fish and Wildlife Service downlisting proposal, published 8 January 2016): https://federalregister.gov/a/2015-32645.

Florida Fish and Wildlife Conservation Commission: www.myfwc.com.

International Union for the Conservation of Nature: www.iucn.org.

Marine Mammal Commission: www.mmc.gov.

Save the Manatee Club: www.savethemanatee.org.

St. Johns River Water Management District (for Brevard County seagrasses): www.sjrwmd.com/itsyourlagoon/2011superbloom.html.

US Fish and Wildlife Service: www.fws.gov.

# Index

Page numbers in *italics* indicate photographs.

geographic information systems, spatial analysis using, 131

gillnet fishing, 116, 126–27

Grand Cul-de-Sac Marin, Guadeloupe, 4, 5, 76

Grand-Terre, Guadeloupe, 1, *2*

gray whales, 121

grazing by manatees, 48, 50

green sea turtles, *59*

Guadeloupe: extinction of manatees of, 1, 2, 4; frigate birds of, *3*; leatherback turtles of, *x*; map of, *2*; reintroduction of manatees in, 4–7

Gulf Coast of Florida, *104*

habitat loss, 98

habitat modification, *22*, 52, 102–3

habitat protection: access to freshwater, 91–93; features, 76–77; human population and, 77, *78*; overview of, 73; quiet waters, 97–98; seagrasses, 75–76; warm-water refuges, 77–79; water quality, 93–97. *See also* warm-water refuges

Haddad, Ken, 131

hairs on manatees, *17*, *18*, 20, 25

harpoons, *23*

Hartley, Wayne, 103

Hartman, Daniel S., 78

H. D. King Power Plant, 86

heat exchangers, counter-current, 20

herbivores, manatees as, 15–16, 24–26, 28–29, 94

hindgut digesters, 26, *27*, 55, 57

hippopotamuses, *27*

Holdren, John, 41

human dimensions viewpoint, 113–14

human population and habitat protection, 77, *78*

hunting of manatees: in developing countries, 116–17; in Florida, *41*; with harpoons, *23*; in Nigeria, 131–32; by subsistence-based societies, 39–41, 42–47, 52, 126; vulnerability and, 23

hurricanes, 110–11

hypothermia, 79

Indian River Lagoon: manatees in, 9–10; threats to, 11–12

Jefferson-Smurfitt Plant, 85

jellyfish, *120*

Kenya, dugongs in, *126*

kidneys of manatees, 20, 93

killer whales, *16*, 64, *65*, 121, 122

Kings Bay, Crystal River, *88*

Laist, David, 120

Lamentin, Guadeloupe, 4

Lauderdale Plant, *79*, 86–87

leatherback turtles, *x*

Leeward Islands of Lesser Antilles, *2*

leopard seals, *16*

Lerebours, Boris, 5

life-history attributes, 21, 23, 34, 67–68, 70, 71

lifespan of manatees, 68, 70, *99*

lips of manatees, *24*, 25

Lynch, Wayne, *63*

MacKenzie, Duncan, 79

Magnin, Hervé, 5

male manatees: in mating herds, 63; sex differences between females, *6*, *17*

management units, 101–2

Manatee Individual Photo-Identification System (MIPS), 75

Manatee Recovery Plan, 77, 89

Manx, 73–74

marine grazers, 59

marine mammals: accumulation of organic chemicals in, 93–94; adaptations of, 16–17; challenges to conservation of, 1–2; flukes of, *33*; overexploitation of, by hunters, 23; reintroduction of, 4–7; species threatened with extinction, 121–22; taxonomic groups of, *16*. *See also* conservation of marine mammals

Marsh, Helene, 71, 120, 122, 128, 129, 131

mating herds, 63, *65*

Meffe, Gary, 123–24

mermaids, manatees as, 41–42

metabolic rate, 19, 79, 81

milfoil, *57*

Miocene epoch, 36

MIPS (Manatee Individual Photo-Identification System), 75

monk seals: Caribbean, 122; Hawaiian, *122*; Mediterranean, 121

Morales, Benjamin, 120

morphological differences in male and female Florida manatees, *6*, *17*

mortality events, 11, 104–5

mothers and calves. *See* female manatees with calves

mouths of manatees, *14*, *18*

Mozambique: dugongs in, 125–27; stamps of, *8*

nails of manatees, *17*

napping, *68*

Nigeria, hunting of manatees in, 131–32

"no action" option, 114–15

noise levels, underwater, 105–6

North American black brants, *59*

nostrils of manatees, *17*

oceans, as under siege, 119

Odell, Dan, 73

oil spills, 93

Oligocene epoch, 36

organochlorine pesticides, 93

Ortiz, Rudy, 79

O'Shea, Tom, 91, 120, 128, 129

osmoregulation, 92–93

osprey, *97*

"paper parks," 130

Parc National de la Guadeloupe, 2, 5, 6–7

Parcs Nationaux de France, 5

Pew Oceans Commission report, 120

pharmaceuticals, as contaminants, 94–95

Pine Island Sound, 56–57, 58

pinnipeds, 16, 35

polar bears, *16*

pollution, waterborne, 93–97

polybrominated diphenyl ethers, 93

polycyclic aromatic hydrocarbons, 93

population: of aboriginal Indians, 40–41; human, and habitat protection, 77, *78*; of manatees, 9, 70–71, 101–2, 136

posture of manatees, *13*

poverty and conservation, 131–32

power plants, warm-water refuges at, 10, 51, *51*, *79*, 85–87, *111*

precautionary approach to risk management, 112–13

predators, 63–64, *65*

pregnancy, *69*

protection for manatees, 43, 45, 46–47, 51. *See also* conservation; habitat protection

Provancha, Jane, 9

quiet waters, 97–98

Ragen, Tim, 120, 122

range: of Florida manatees, 3, *4*, 18–19, *32*; of West Indian manatees, 127–28

red mangrove seedlings, *57*

red tides, 95–96, *97*, 105

regulatory tools for conservation, 129–30

reintroduction of manatees in Guadeloupe, 4–7

reproduction: disruption or impairment of, 95; estrous herds, *65*; injury and, 105, *107*; life-history attributes and, 21, 23, 67–68; population growth rate and, 70–71; scramble competition, 63; sex play, *69*

rhinoceros, black, *27*

Ricklefs, Robert E., 21

right whales, *16*

Riviera Plant, 51, *79*, 86

Rose, Pat, 120

ruminants, 26, *27*, 57

Runge, Mike, 70

satellite tags, *124*

Save the Manatee Club, 101

science and conservation, 122–23

Scolardi, Kerri, 9

scramble competition, 63

seagrasses: in Brevard County, 11, 58, 76; decline of, 75–76, 96–97; in diet, *11*, *36*, 56–59; evolution of, 36; scarring of, 106; types of, *57*

seals, 16, *16*, 121, *121*, 122, *122*

sea nettles, *120*

sea otters, 5, *16*

senses used by marine mammals, 20, 61, 63

sex determination Florida manatees, *6*, *17*

sexual maturity, 68, 70

shark fins, 125–26

sharks, 64, *65*, *125*

signage, *108*

*Silent Sirens* (documentary), 15